N: 978-981-14-1658-3

: 978-981-14-1657-6

TROGEN REMOVAL PROCESSES
R WASTEWATER TREATMENT

or:

is Hoseinzadeh

Nitrogen Removal Processes for Wastewater Treatment

Edited by

Edris Hoseinzadeh

Department of Environmental Health Engineering,
Social Determinants of Health Research Center,
Saveh University of Medical Sciences, Saveh, Iran

Nitrogen Removal Processes for Wastewater Treatment

Editor: Edris Hoseinzadeh

ISBN (Online): 978-981-14-1658-3

ISBN (Print): 978-981-14-1657-6

© 2019, Bentham eBooks imprint.

Published by Bentham Science Publishers Pte. Ltd. Singapore. All Rights Reserved.

need for a court order if at any point you breach any terms of this License Agreement. In no event will any delay or failure by Bentham Science Publishers in enforcing your compliance with this License Agreement constitute a waiver of any of its rights.

3. You acknowledge that you have read this License Agreement, and agree to be bound by its terms and conditions. To the extent that any other terms and conditions presented on any website of Bentham Science Publishers conflict with, or are inconsistent with, the terms and conditions set out in this License Agreement, you acknowledge that the terms and conditions set out in this License Agreement shall prevail.

Bentham Science Publishers Pte. Ltd.
80 Robinson Road #02-00
Singapore 068898
Singapore
Email: subscriptions@benthamscience.net

CONTENTS

FOREWORD

Since the ecological and environmental monitoring and protection has been widely addressed and emphasized by academic research and public, the remediation and removal for the pollutants became a critical subject for the Industry, Official and Academy. The state-of-the art technology needs to be comprehensively surveyed covering a wide spectrum of related areas. This book is aimed to fulfill this goal and hopefully will benefit the student, teaching staff, engineers, staffs in the civil corporation and also the public.

This book is divided into 5 chapters and an introduction. In the introduction section, Nitrogen compounds and their toxicity and pollution in the water environment are first addressed and the necessity to remove them from the water body is emphasized. Chapter 1 discusses conventional method and technology of Nitrogen removal from water. Chapter 2 covers conventional Biological method for nitrogen removal. The next three chapters are the essence of this book and maybe the most interesting parts for the readers, in Chapter 3 the authors introduce food industry wastewater treatment from nitrogen compounds. In Chapters 4 and 5 a new approach towards environmental pollution remediation biotechnology by different types of Bio Electrochemical Systems (BES) is presented. The authors exert our effort to make this book as an important reference and manual guide for the readers interested in the field.

Dr. Chiang Wei
The Experimental Forest,
College of Bio-Resources and Agriculture,
National Taiwan University
Taiwan

PREFACE

As we know, most of the human activities that require the use of water produce wastewater. As the total demand for water grows, the quantity of wastewater has increased and its overall pollution load is continuously increasing worldwide. One of the pollutants is the nitrogen compound in wastewater. Nowadays, attention to nitrogen removal in wastewater treatment processes has increased. The discharge of nitrogen compounds plus other nutrients into the water bodies leads to water pollution and the growth of algae and aquatic plants. The ground water and surface water supplies are at risk of pollution by these compounds in many areas of the world. From another perspective, water demand is greater than the amount supplied by rain and snowmelt. Therefore, water conservation, wastewater recycling, and reuse are becoming more important. Removal of nitrogen from wastewaters can be highly focused as it can be removed simply and effectively. To successfully remove nitrogen from wastewater, we must have the basic knowledge. We have, therefore, found the description of nitrogen removal techniques in wastewater treatment processes that can be useful and interesting for readers.

The introduction section of chapter 1 gives a very brief overview on the nitrogen compounds and its toxicity, highlighting pollution in the water environment and the necessity to remove this compound from water. Chapter 1 discusses conventional methods and the technology of nitrogen removal from water. Chapter 2 present a conventional biological method for nitrogen removal in wastewater treatment plan. The next three chapters are the essential parts of this book and maybe the most interesting parts to the readers. The authors introduce the role of food industry in wastewater treatment for nitrogen compounds in chapter 3. In chapters 4 and 5, a new biotechnological approach toward eradicating environmental pollution by different types of Bio Electrochemical Systems (BES) is prepared. Although the present eBook is not design for process engineers specially, but by studying this book, the reader would have an overview of the risk related nitrogen compounds to the environment, along with conventional and new techniques to remove them from wastewater. The principal of the proposed method along with practical information and effective factors on the each process is given that makes this eBook a very useful document in the area of nitrogen removal process. In general, this book is very easy to read and understand. This book offers major developments in recent research to bring nitrogen removal processes in to practice. It provides a unique, holistic perspective on the basic and applied research issues regarding the application of nitrogen removal processes in wastewater treatment. The reader finds up-to-date information and solutions of nitrogen removal processes from many leading experts in the field.

This eBook is not free of limitations. Though we tried our best to make the book free of errors and limitations, we apologize if any error is found. Please feel free to contact us and reflect your comments to improve this eBook in next editions. I would like to thank my family (my patient wife and lovely son, Azhwan) for their time and special thanks to my colleagues, teachers and scientists for contributing.

Edris Hoseinzadeh
Department of Environmental Health Engineering,
Social Determinants of Health Research Center,
Saveh University of Medical Sciences,
Saveh,
Iran

ACKNOWLEDGEMENTS

I would like to thank the all colleagues who have contributed to the development of present edition. Students, classroom instructors, and reviewers provided constructive comments and suggestions that were most helpful during the revision. I also received several comments from readers of the related publications. Their encouraging comments are appreciated. I am grateful to Dr. Chiang Wei, The Experimental Forest, College of Bio-Resources and Agriculture, National Taiwan University for preparing the preface and his comments.

CONFLICT OF INTEREST

The authors declare no conflict of interest.

List of Contributors

Saeb Ahmadi	Tarbiat Modares University, Tehran, Iran
Edris Hosseinzade	Saveh University of Medical Sciences, Tehran, Iran
Ashkan Sami Vand	Shiraz University, Shiraz, Iran
Hooshyar Hossini	Kermanshah University of Medical Sciences, Kermanshah, Iran
Ghazaleh Mirbolouki Tochaei	Guilan University of Medical Sciences, Rasht, Iran
Mehrdad Farrokhi	University of Social Welfare and Rehabilitation Sciences, Tehran, Iran
Mehrdad Moslemzadeh	Guilan University of Medical Sciences, Rasht, Iran
Saeid Ildari	Guilan University of Medical Sciences, Rasht, Iran
Mostafa Mahdavianpour	Abadan School of Medical Sciences, Abadan, Iran
Mahdi Farzadkia	Iran University of Medical Sciences, Tehran, Iran
Saman Hoseinzadeh	Iran University of Sciences and Technology, Tehran, Iran
Khadijeh Jafari	Hormozgan University of Medical Sciences, Bandar Abbas, Iran
Mahshid Loloei	Kerman University of Medical Sciences, Kerman, Iran
Reza Shokuhi	Hamadan University of Medical Sciences, Hamadan, Iran
Bahram Kamarei	Lorestan University of Medical Sciences, Khorramabad, Iran
Reza Barati Rashvanlou	Iran University of Medical Sciences, Tehran, Iran
Mohammad Mahdi Shadman	Tarbiat Modares University, Tehran, Iran
Omid Rahmanian	Hormozgan University of Medical Sciences, Bandar Abbas, Iran
Mohammad Amin Mirnasab	Shiraz University of Medical Sciences, Shiraz, Iran
Ghorban Asgari	Hamadan University of Medical Sciences, Hamadan, Iran
Ayub Ebadi Fathabad	Urmia University, Urmia, Iran
Sakine Shokoohyan	Hormozgan University of Medical Sciences, Bandar Abbas, Iran

CHAPTER 1

Conventional Methods and Techniques of Nitrogen Removal from Water

Saeb Ahmadi[1], Saman Hoseinzadeh[2], Mohammad Mahdi Shadman[1], Omid Rahmanian[3] and Khadijeh Jafari[4,*]

[1] *Department of Chemical Engineering, Tarbiat Modares University, Tehran, Iran*

[2] *Department of Civil Engineering, Iran University of Sciences and Technology, Tehran, Iran*

[3] *Department of Environmental Health Engineering, School of Health, Hormozgan University of Medical Sciences, Bandar Abbas, Iran*

[4] *Department of Environmental Health Engineering, School of Health, Hormozgan University of Medical Sciences, Bandar Abbas, Iran*

Abstract: Nitrogen-containing compounds are among the pollutants that can cause serious environmental hazards. One of these hazards is nutrients enrichment of rivers that can result in eutrophication, decreased water quality, and potential health hazards for humans and animals, when released in the environment. Nitrate removal methods can be generally classified into physical, biological, and chemical reduction methods. The most commonly used methods in this regard are biological denitrification, ion exchange, electrodialysis, reverse osmosis, chemical denitrification, adsorption, electrocoagulation, nanotechnology, and redox reaction. The first four methods have been used in the industry. Biological denitrification is an effective method because of the conversion of nitrate into N_2 gas and the absence of secondary pollutant production. However, it is not widely used in the removal of nitrate from drinking water sources and underground water due to microbial contamination and rather is mostly used for wastewater treatment. The purpose of this study is to present a brief introduction on the use of physiochemical methods for the removal of nitrate from water and wastewater.

Keywords: Adsorption, Electrodialysis, Electrochemical Reduction (ER), Ion Exchange, Nitrate, Reverse Osmosis, Redox (Oxidation-Reduction) Reactions.

INTRODUCTION

Water is an essential compound in human life because metabolic and synthetic mechanisms are closely interconnected to the specific properties of water.

* **Corresponding author Khadijeh Jafari:** Department of Environmental Health Engineering, Student Research Committee, Department of Environmental Health Engineering, School of Health, Hormozgan University of Medical Sciences, Bandarabbas, Iran; Tel/Fax: +989140703556; E-mail: k.jafary.71@gmail.com

Transfer of nutrients to cells and interaction with the environment are impossible without water. However, water resources are limited such that only 2.66% of the total water resources of the world including groundwater, lakes, rivers, arctic ice, and glaciers are fresh waters. In addition, just a small portion of fresh waters (about 0.6%) can be consumed as drinking water. Therefore, it is imperative to treat water resources and wastewaters, efficiently [1]. The high tendency of humans toward urbanization, industrialization, and agricultural activities has disposed different pollutants to the environment. Nitrogen is among the essential nutrients for the survival of living creatures. The main source of this element is the atmosphere of the earth. Nitrogen exists in different forms, such as (NO_3^-), ammonia, ammonium, and nitrogen gas (N_2), which can transform into each other under different procedures and conditions [2, 3]. To facilitate understanding of nitrogen transformations, the nitrogen cycle is depicted in Fig. (**1**).

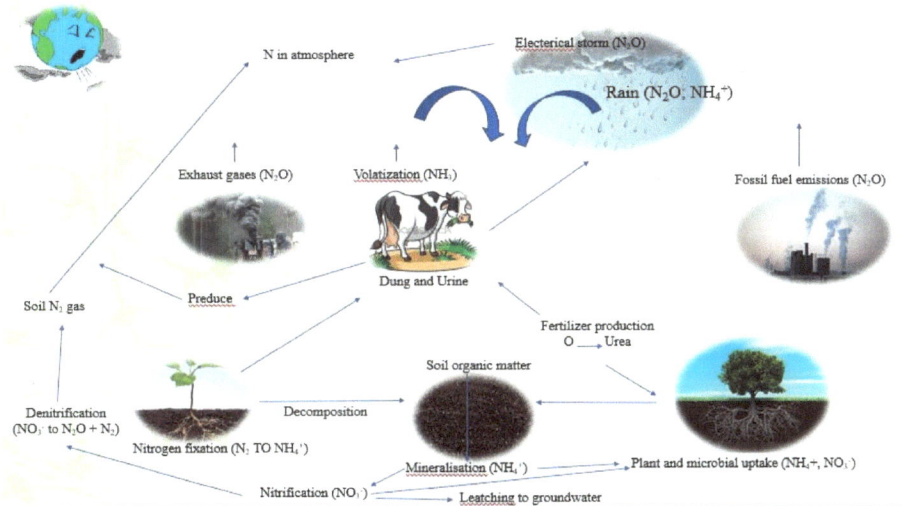

Fig. (1). The nitrogen cycle in the environment.

Nitrogen-containing compounds are among the pollutants that can cause serious environmental hazards. One of these hazards is nutrients enrichment of rivers that can result in eutrophication, declined water quality, and potential health hazards for humans and animals, upon emission to the environment [4]. One of the most notable nitrogen compounds for environmentalists is nitrate. To date, many researchers have attempted to find appropriate solutions for the removal of nitrate from water resources, aquaculture ponds, aquariums, and industrial wastewaters. Pollution of groundwater resources by nitrate is one of the main environmental concerns. In fact, nitrate is known globally as one of the most common chemical pollutants of groundwater. Chemical fertilizers, animal wastes, and inappropriate disposal of human and animal wastes are the main sources of nitrate emission to groundwater [5 - 7]. Although nitrate is not a direct threat to the health status of humans and animals, it can potentially be converted to nitrite in the gastric system or reduced into nitrosamine compounds, which endanger the global health and lead to different kinds of gastrointestinal and liver cancers [8 - 11]. In children younger than 6 years, nitrate can be converted into nitrite and bind to human hemoglobin, leading to methemoglobin formation. Methemoglobin formation reduces oxygen supply to body tissues and causes blood darkening. This illness is called methemoglobinemia or the blue baby syndrome [12]. According to the guidelines of WHO and EPA, the maximum permitted concentration of nitrate for public water systems is 10 mg L^{-1} and 50 mg L^{-1} based on the nitrogen and nitrate contents, respectively [10, 13]. In the remainder of this paper, nitrate is described as a water pollutant and the related issues are addressed. Then, the common physical and chemical methods that can help to eliminate nitrate from water resources are discussed.

Physical and Chemical Methods of Nitrogen Removal from Water and Wastewater

Conventional water treatment methods such as filtration have no significant effect on nitrate removal because nitrate is stable and highly soluble with low potential for absorption or precipitation. The increasing amount of nitrate in underground water resources has become one of the main environmental problems. Therefore, many different biological and physicochemical methods including reverse osmosis, ion exchange, electrodialysis, biological denitrification, application of nanotechnology [14, 15], and chemical denitrification have been studied by many researchers for the removal of nitrogen components. Most of the physicochemical methods have high yields, but they usually produce a concentrate nitrate wastewater and require pre-treatment and post-treatment processes. Each of these methods has their own advantages and disadvantages as described in the following.

Ion Exchange

The process of ion exchange due to low cost and high flexibility is one of the effective methods for nitrate removal. Ion exchange is a reversible ion process that is used to remove pollutants and water hardness [16]. The process efficiency is determined by two factors: the concentration and attraction of ions in the solution. Ion exchange resin is an insoluble matrix usually made of a polymer that removes ions to purify or eliminate water hardness [17].

In anion exchange resins, nitrate usually replaces chloride and sulfate ions. The problem of these resins is their greater tendency to absorb sulfate instead of nitrate in the solution, so the process efficiency is usually decreased and the concentration of nitrate and sulfate in the solution should be considered. Sodium chloride and sodium bicarbonate solutions are commonly used to recover resins and remove adsorbed nitrate ions. In order to maintain the efficiency of the system, when the system is in service mode for a long time, the produced wastewater must be discarded because of the organic matter and salts in the wastewater cause resin fouling [18].

In the ion exchange process, the water enters the ion exchange chamber after passing through the pretreatment process, and the nitrate is absorbed with resin sites by displacement of chloride ions. This absorption mechanism is similar to eliminating the water hardness in water softeners [19]. Eq. 1 presents the displacement mechanism of nitrate onto an SBA resin in the chloride. Fig. (**2**) shows the schematic of the process.

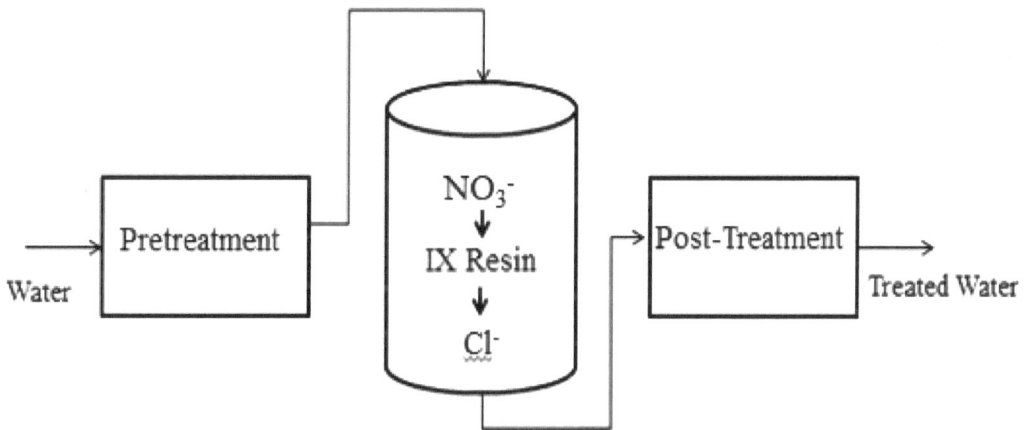

Fig. (2). Ion exchange schematic.

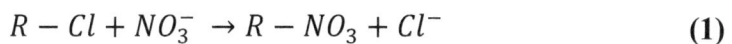

$$R - Cl + NO_3^- \rightarrow R - NO_3 + Cl^- \tag{1}$$

Many researchers have examined the ion exchange resin process for nitrate removal. Samatya *et al.* used a strong-base ionic exchange resin for the removal of nitrate in groundwater and reported a total capacity of 157 mg/g and a yield of about 81%. They also used a 0.6 M NaCl solution to recover the resin [20].

Milmile *et al.* used the ion exchange in Dion NSRR resin to remove nitrate from the solution. The absorption capacity of the system was reported to be about 119 mg/g as per Langmuir isotherm [21].

Ion exchange resins are also used to remove ammonia from wastewater.

Clinoptilolite is the most common zeolite. The mechanism for removing ammonia in this silicon-rich zeolite is based on the exchange of cation and adsorption in the pores of the zeolite. At higher temperatures, the ammonia removal efficiency is higher. Solutions with a high concentration of calcium ions can be used for regeneration of zeolite [22].

Ion exchange resins are divided into two types of acrylic and styrene, which consist of an aliphatic and aromatic matrix, respectively. The advantage of these resins is that they can easily be recovered using NaCl, KCl, and NaOH solutions. The presence of anions such as phosphate, sulfate, and bicarbonate reduces the efficiency of nitrate removal in anion exchange resins. The affinity sequence of the anion exchange resins is as follows [23]:

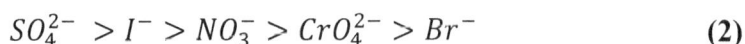

$$SO_4^{2-} > I^- > NO_3^- > CrO_4^{2-} > Br^- \tag{2}$$

Anion exchange resins are used to solve the problem of anions competition. One of the resins selected for nitrate is macroporous styrene strong-base anion (SBA), which is commercially known as the A520E Purolite [23].

The advantages of ion exchange systems include the possibility of simultaneous elimination of several pollutants, selective nitrate removal, economical process, small and large-scale applications, and automation. The probability of nitrate dumping, the need to adjust the pH, resin fouling, and producing wastewater are some disadvantages of these systems [24].

Reverse Osmosis (RO)

In the reverse osmosis (RO) method, the nitrate is removed from the water with a membrane using osmotic pressure [18]. Moreover, the RO process can simultaneously remove several pollutants (*e.g.*, nitrate, chloride and fluoride, and a variety of salts). The mechanism of this process is that the water passes through a semi-permeable membrane while the membrane prevents the passage of contaminants [25].

In the RO process, nitrate and other pollutants are in the pressure section, and pure water is transferred to another section through a permeable membrane. This process occurs due to the pressure difference between the two sides of the membrane [26]. To remove nitrate, a composite thin membrane made of cellulose is usually used. Using these membranes, the nitrate concentration can be reduced by 60 to 95%. The water nitrate content more than 30 mg/L reduces the RO process efficiency [27]. Parameters of nitrate concentration, water pretreatment process, membrane material, osmotic pressure, temperature, and drainage of concentrated wastewater must be taken into account in this process [28].

The advantages of this process include the production of very high-quality water, the simultaneous elimination of different sources, the desalination of water, application in small systems, and rapid launch. The disadvantages of the RO process, on the other hand, include the need for pretreatment process and membrane fouling problems. The processes such as filtration, flocculation, pH adjustment, and anti-scaling are used for pretreatment [29].

The membranes are made of materials such as polyamides, cellulose acetate, and composite materials. The most important application of this process is in water desalination. Darbi *et al.* compared three methods of RO, ion exchange, and biological processes in terms of nitrate removal from water and found that the low nitrate removal efficiency was related to the RO process [30].

Hayernen *et al.* showed that using the RO process could remove 90% of nitrate [31]. Richards *et al.* investigated the effect of pH on RO process performance in nitrate removal and showed that pH has no effect on nitrate removal efficiency [32]. Fig. (**3**) gives a schematic illustration of the RO process.

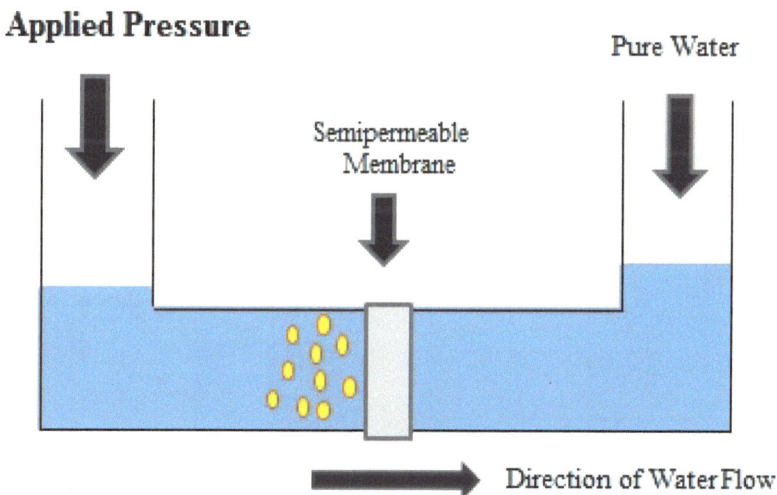

Fig. (3). Reverse osmosis schematic.

Electrodialysis

The electrodialysis (ED) method has a great application in nitrate removal from water because of the low consumption of energy and chemicals, good selectivity for nitrate ions, and low effluent production [30]. The basis of ED systems is using ion exchange membranes and applying an electrochemical potential, as the driving force. ED is performed by passing a direct electric current through ion-selective membranes and nitrates while other ions being transferred from the dilute solution to a permeate solution [16]. The wastewater of the process is generated through a concentrated solution of ions and nitrates. The ED system consists of anion exchange membrane and cation exchange membrane placed between two electrodes. By applying the electrical potential difference, the anions in a low-density solution pass through anionic exchange membrane and move toward the positive charge anode. Cations also pass through the cation exchange membrane toward the negative charge cathode. ED systems perform the water purification process by removing charged ions from the dilute flow and increasing the concentration of ions in the permeate flow [33]. To remove nitrate or ions, the semipermeable membrane should be selected based on target ions [16]. It is over 50 years from the first use of ED. In some industrial ED systems, there are 100 to 200 pairs of membrane cells between electrodes [30, 34]. ED is an efficient approach such that 94.1% nitrate removal per inlet current was reported through ED of 100 mg L^{-1} nitrate solution in the presence of organic compounds [35]. The pH range of 3 to 5 can result in decreasing the concentration of nitrate from 173.8 to 4-5 mg L^{-1} during 42 min ED processing, which demonstrates that ED can be affected by pH. The schematic of the ED process is shown in Fig. (**4**). Compared to other membrane methods, ED has a higher efficiency.

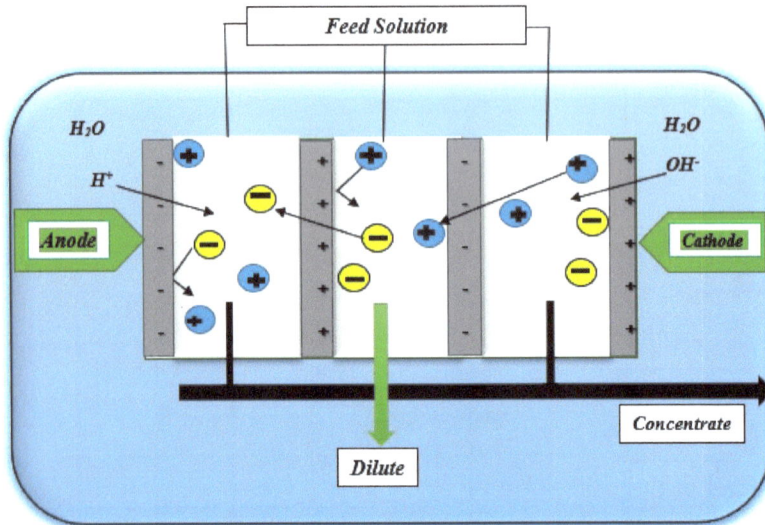

Fig. (4). Electrodialysis schematic.

The reversal ED is used to reduce the problem of membrane fouling. In this process, the polarity of the electrodes is changed with changing the flow direction to change the direction of the ions. By changing the direction of flow, the ions pass through the membrane in the opposite direction. The nitrate removal efficiency in this process is higher than conventional ED processes [36].

Elmidaouni *et al.* used 5 types of anion exchange membranes and one type of cation exchange membrane for nitrate removal. They reported a decrease in nitrate concentration from 113 mg/L to 30 mg/L using 1 h ED operation. They also succeeded in reducing the amount of nitrate in groundwater with a concentration of 90 mg/L and salinity of 800 mg using electrodialysis method [37]. Banasiak *et al.* studied the removal of nitrate from organic materials by electrodialysis and found a removal efficiency of about 94.1% [35]. The effect of pH on the nitrate removal using the ED method was investigated by Abou-Shady *et al.* [38], who showed that the optimal pH value for this process was between 3 and 5.

The superiority of ED over reverse osmosis, as two common methods, is that ED can be adopted for treating different solutions with high levels of ions because this technique is independent of osmosis pressure. Moreover, lack of precipitation on ED membranes obviates the need for pre-treatment. In addition, the high quality of the output water, low operational cost, the long life of the membrane, the possibility of selective elimination of various pollutants, simultaneous elimination of multi-pollutants, and lack of requiring membrane activation (like ion exchange methods) are the advantages of ED. On the other hand, the inefficiency of ED for the elimination of fine particles and organic materials, need for diluting concentrated solutions after the entrance to the system, and an increase in investment costs due to energy consumption and a greater number of large membranes are the disadvantages of ED [39].

Physico-Chemical Removal of Ammonia

Recently, methods like ammonia air and steam stripping, ammonia vacuum distillation, and ammonia adsorption have been developed for the removal of ammonia from wastewater. In the following, some studies conducted in this regard are presented:

Ammonia Air and Steam Stripping

It is possible to remove ammonia from concentrate wastewater solutions using air and steam stripping. The process efficiency depends on temperature, pH, and operating parameters such as liquid flow rate, the height of the filled tower, and so on. To adjust the pH between 10.8 and 11.5, a small amount of lime is added to the sample. In this regard, air stripping has found a greater number of applications than steam stripping [40]. Moreover, flash distillation can be used to recover ammonia. Fig. (**5**) presents the process schematically. The ammonia content can be declined from 1000 mg/L to 100 mg/L using this process at 60°C and pH = 9.5 [41].

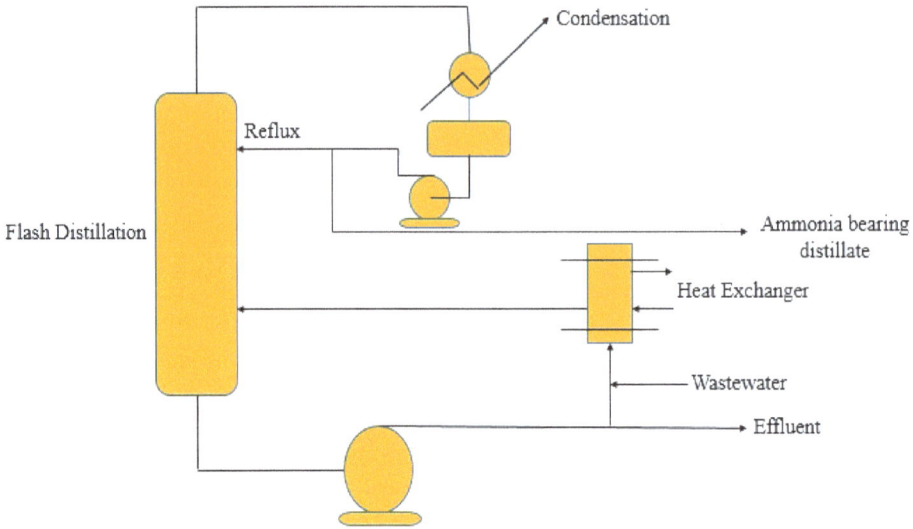

Fig. (5). Flash distillation for ammonia recovery schematic [22].

The Deposition of Ammonia in The Form of Struvite

Precipitation of NH_4^+ as magnesium ammonium phosphate (MAP), also called struvite, is another method for wastewater treatment [42]. The general reaction of this process is as follows:

$$Mg^{2+} + PO_4^{3-} + NH_4^+ + 6H_2O \leftrightarrow MgNH_4PO_4.6H_2O \downarrow \qquad (3)$$

The MAP is easily separated from water due to its low solubility in water, so it is possible to separate ammonia by deposition in the form of an MAP. This process is controlled by temperature, pH and impurities effects. In this process, the crystalline deposits are formed through nucleation [22].

AMMONIA AND CHLORAMINES REMOVAL BY ACTIVATED CARBON

Activated carbon can be used to adsorb ammonia. However, the adsorption performance in the soluble phase is poor due to its strong affinity with water. To overcome this problem, bamboo charcoal carbonized at 400°C can be used to adsorb ammonia. Moreover, active carbon has a highly beneficial effect on chloramine elimination. Chloramine is produced as a result of the reaction of chlorine and its by-products with water. Although chloramines like chlorine have disinfectant properties, removing them from the water is essential due to their negative effects on the odor and taste of water [43].

The reactions of the formation of chloramine are as follows [22]:

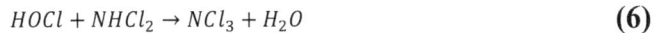

$$HOCl + NH_3 \rightarrow NH_2Cl + H_2O \qquad \textbf{(4)}$$

$$HOCl + NH_2Cl \rightarrow NHCl_2 + H_2O \qquad \textbf{(5)}$$

$$HOCl + NHCl_2 \rightarrow NCl_3 + H_2O \qquad \textbf{(6)}$$

The formation of chloramines depends on the pH and molar ratio of $Cl_2:NH_4^+$. For example, dichloramine is a dominant sample at pH 7-8 and molar ratios of $Cl_2:NH_4^+ > 1.8$, but at the ratio of $Cl_2:NH_4^+ < 0.7$, mono chloramine is prevailing [44].

Nanotechnology

Nanotechnology, as an advanced water treatment method, is the engineering and skill of applying nano-sized (1 to 100 nm) materials [45]. Each nano-based method is efficient for the removal of one specific type of pollutant. In addition, a combination of physical and chemical approaches and nanotechnology can be used for the advanced treatment of groundwater and wastewater [46, 47]. Among nano-based methods, nanofiltration is more efficient in the elimination of various ions, such as nitrate, from groundwater and wastewater [48]. In a study, the performance of four different types of nanofiltration (NF) membranes was compared and found that four membranes were efficient for the removal of aqueous nitrate [49]. To produce nanoparticles (NPs), zero-valent metals, such as Fe, Zn, and Sn [32], and/or paramagnetic particles, such as Fe_3O_4 and Fe_2O_3, can be used. Because of the small size of NPs and their large surface areas, NPs can provide high efficiencies for the removal of pollutants [15]. Alumina NPs show a high efficiency in the removal of nitrate from water. Also, they are recommended for off-site water treatment services [14]. In addition to alumina NPs, nZVI have been used for nitrate removal since they are safe, eco-friendly, low-cost and convenient. Furthermore, the reaction of nitrate with nZVI just gives nitrogen and ammonia and does not lead to the generation of any hazardous compound [50]. One of the most interesting recent research fields is the application of zero valent iron NPs (nZVI) for removal of pollutants from aqueous environments. Since nZVI particles are small, applying pressure or centrifugal force to them can give a slurry solution, transmit them to the polluted zones, and treat the pollutants *in situ*

[51, 52]. The main challenge of using nZVI particles is their strong tendency toward oxidation, which can be alleviated by employing stabilizing agents [53]. To stabilize nZVI particles, water-soluble polysaccharides can be used owing to their low prices and environmental compatibility [51]. Among such polysaccharides, starch and cellulose are more common [54].

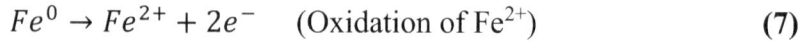

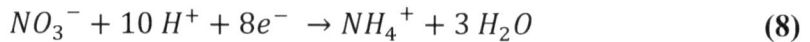

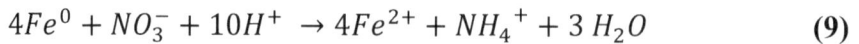

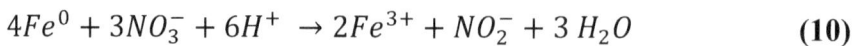

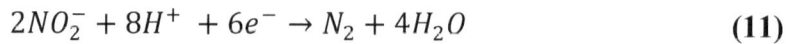

$$Fe^0 \rightarrow Fe^{2+} + 2e^- \quad \text{(Oxidation of Fe}^{2+}\text{)} \tag{7}$$

$$NO_3^- + 10\,H^+ + 8e^- \rightarrow NH_4^+ + 3\,H_2O \tag{8}$$

$$4Fe^0 + NO_3^- + 10H^+ \rightarrow 4Fe^{2+} + NH_4^+ + 3\,H_2O \tag{9}$$

$$4Fe^0 + 3NO_3^- + 6H^+ \rightarrow 2Fe^{3+} + NO_2^- + 3\,H_2O \tag{10}$$

$$2NO_2^- + 8H^+ + 6e^- \rightarrow N_2 + 4H_2O \tag{11}$$

Although nanotechnology is advantageous for treating groundwater and wastewater, it is toxic for some species and has some adverse effects on aquatic species. Moreover, NPs can accumulate in fatty tissues of living organisms and transfer gradually *via* the food chain,. Therefore, the adverse impacts of applying NPs on the environment and health of different species should be assessed when the profits of using NPs are considered [55].

Electrochemical Processes

Electrochemical approaches having applications in nitrate removal include the new methods of electrical reduction (ER), electrocoagulation (EC) and ED. ER and EC techniques, nitrate is transformed into a non-toxic and non-hazardous form. However, ED is a physical method and does not alter the chemical form of nitrate [56]. In electrochemical methods, there are two electrodes; *i.e.*, an anode and a cathode. On the surface of the cathode, nitrate reacts with water and converts to nitrite. Consequently, the reaction of the produced nitrite and water gives nitrogen gas and ammonia. These cathodic reactions require 8 electrons that can be supplied through hydrolysis of water at the anode. Nitrite and ammonia are the secondary products of this process and decline the efficiency of electrochemical nitrate removal [57, 58]. Some disadvantages of electrochemical methods are high consumption of energy, noticeable variation in their results, and the necessity of continuous maintenance. In addition, similar to the two common nitrate removal methods of RO and ion exchange, the electrochemical method cannot be implemented for the removal of nitrate ions selectively and requires the continuous replacement of electrode. In this regard, in electrochemical methods, unlike these two common methods (*i.e.*, RO and IE), the water treatment trends have shifted toward using electrochemical techniques [8, 59].

Electrochemical approaches are associated with several advantageous including selective nitrate removal, not requiring chemical materials for coagulation procedure, low sludge production, high efficiency, low cost, and eco-friendly [57]. Moreover, research findings have indicated that electrochemical processes can convert organic pollutants to non-hazardous gases such as carbon dioxide and nitrogen [60 - 62]. Some parameters that influence the efficiency of electro-chemical methods include the homogenous distribution of current, electrode stability, electrode's material, pH, the applied voltage, and high mass transfer [63, 64]. The extent of nitrate removal is directly and inversely related to the density of input current and concentration of NaCl, respectively [57]. The reactions involved in the removal of nitrate ions through electrochemical methods are listed in Table 1.

Table 1. Reaction of nitrogen removal by electrochemical method.

Process	Reaction
Electrolysis of water at cathode	$2H_2O + 2e^- \rightarrow H_2 + 2OH^-$
Anodic electrolysis of water	$4OH^- \rightarrow O_2 + 2H_2O + 4e^-$
Reactions of nitrate ion and water molecules	$NO_3^- + H_2O + 2e^- \rightarrow NO_2^- + 2OH^-$ $NO_3^- + 3H_2O + 5e^- \rightarrow 0.5\,N_2 + 6OH^-$ $NO_3^- + 6\,H_2O + 8e^- \rightarrow NH_3 + 7OH^-$
Reactions of nitrite ion and water molecules	$NO_2^- + 2H_2O + 3e^- \rightarrow 0.5N_2 + 4OH^-$ $NO_2^- + 5H_2O + 6e^- \rightarrow NH_3 + 7OH$ $NO_2^- + 4H_2O + 4e^- \rightarrow NH_2OH + 5OH$
Reduction of nitrate to produce ammonia	$NO_3^- + 2H_2O \rightarrow NH_3 + 2O_2 + OH^-$
NaHCO₃ added to maintain pH of electrolyte	$NaNO_3 + NaHCO_3 + H_2O \rightarrow NH_3 + 2O_2 + Na_2CO_3$
Chlorine formed in anodic electrolysis	$2Cl^- \rightarrow Cl_2 + 2e^-$
Reaction of chlorine and water molecules	$Cl_2 + H_2O \rightarrow HOCl + H^+ + Cl^-$
Reaction of nitrite and HOCl ions	$NO_2^- + HOCl \rightarrow NO3^- + Cl^- + H_2O$
Reaction of ammonium and HOCl	$2NH_4^+ + 3HOCl \rightarrow N_2 + 5H^+ + 3Cl^- + 3H_2O$

Electrochemical Reduction (ER)

The mechanism of ER relies on electron transfer and hydrogen reduction. ER of nitrate ions to non-hazardous nitrogen has been conducted for several decades because of its good operational conditions and high efficiency. For example, the ER method can lower the nitrate content of the textile industry effluents about 85% within 3 h of removal processing [65]. The mechanism of ER reactions highly depends on the cathode's material, cathodic potential, and pH of the working solution. Typically, the major products of ER processes are nitrogen gas, nitrite, and ammonia [66 - 68]. One of the important and essential factors affecting the performance of reductive electrochemical systems is the type of material employed as the electrode, which can be either metallic or non-metallic [68, 69]. The most important non-metallic electrodes for ER systems are boron-doped diamond (BDD) and graphitic electrodes. BDD has a very high efficiency in pollutant removal. However, the costs of these electrodes are high, which restricts their practical applications [70 - 72]. Some of the other parameters that should be concerned as effective parameters on the performance of ER techniques are potential [73], pH [65], the electrolyte material [74], temperature, and the reactor type [75, 76].

Electrocoagulation (EC) Processes

Since the use of chemical coagulants to the outlined physiochemical methods, a great sludge mass containing pollutants and toxic materials is generated [77]. The efforts of researchers about utilizing electricity to treat water and wastewater have developed noticeably in recent years [78 - 80]. EC is an electrolysis reaction that requires an external electrical source to apply a suitable electrical potential between two or more electrodes (Fe, Al, or Ti) and carry out some electrochemical reactions at the interface of electrode-solution. In this method, the pollutant removal agents (*i.e.*, aluminum and iron hydroxide clots) are produced by applying an electrical current on a set of floating electrode planes. Therefore, no chemical coagulant is required for this purpose [81, 82]. When iron or aluminum is used as the electrodes of EC systems, Fe^{III} and Al^{III} are generated. The produced metal ions can react with hydroxyl ions to generate poly hydroxide ions or metal hydroxides [83]. In summary, EC is a combination of the dissolution, coagulation, and flocculation mechanisms. In the first step, the coagulant agent is produced using anodic oxidation. The second step involves destabilization of the pollutant. In the third step, pollutants are trapped by the resultant aggregate-flocs [84]. Reduction of nitrate to nitrogen gas is possible by the EC method through the anodic production of hydroxide ion from dissolved iron ions [85]. In this respect, the findings of several studies have shown that aluminum electrodes are more appropriate for F^- elimination from water rather than for nitrate removal [86, 87]. EC weakly depends on the type of wastewater and can continue the removal process as long as the energy source of the electrodes is supplied. Also, selectivity, removing a wide spectrum of pollutants, ease of application, the simplicity of equipment, and removal of the pollutants in the short time are advantages to the EC processes [88]. Besides, EC can provide higher efficiency, lower sludge formation, and conversion of pollutants to safe material compared with the conventional chemical methods [89, 90]. The most considerable disadvantages of EC are the regular replacement of the electrodes owing to their dissolution, high electricity cost, anode explosion, and formation of gelatinized hydroxides that might tend to dissolve. These issues can be solved by reflowing the solution and using solar energy [91]. Nitrate removal by EC has a relatively good efficiency. Also, there is a negligible difference between the amount of pollutant removed from distilled and tap water when EC is performed [91]. To optimize EC processes, pH, type of electrode, current density, energy consumption, and nitrate removal mechanism can be regarded as the parameters that might affect the efficiency of EC [92]. Many studies have focused on removing nitrate from groundwater and wastewater by EC and have reported its high efficiencies [92, 93].

Redox (Oxidation-Reduction) Reactions

Redox reactions are among the important reactions in chemical water treatment, particularly in the chemical treatment of wastewater. In these reactions, an ion, atom or molecule oxidizes when it loses an electron and reduces when it gains an electron. Advanced oxidative processes (AOPs) are water treatment approaches that are performed *via* oxidation of different compounds over the temperature and pressure range of water. AOPs rely on producing a sufficient number of OH radicals to degrade pollutant molecules [94]. AOPs, such as homogenous and heterogeneous photocatalytic processes, photochemical degradation, sonolysis, ozonation, and chemical oxidation processes, have been successfully performed to decline the level of various pollutants in water and wastewater samples [95 - 97]. Fenton and wet oxidation methods are two other examples of AOPs. During all AOPs, organic materials are decomposed into minerals [98]. One of the most common areas of applying AOPs is the elimination of nitrogen-containing compounds from the effluents of the petrochemical industries that produce urea and ammonia [99]. Modification of the Fenton method by nZVI particles has been proved to be effective for the removal of nitrate under optimum conditions. Using iron compounds at low pH's and the presence of hydrogen peroxide can convey a positive impact on decomposition and reduction of nitrate content of groundwater. However, it should be considered that oxidative processes might result in recycling of some new pollutants, specifically organic compounds [100]. Treatment of chemical wastewaters by AOPs converts pollutants to CO_2, minerals, and water or transforms pollutants into environmentally non-hazardous compounds. In addition, non-biodegradable organic materials can be partially decomposed by AOPs and later degrade completely by conjugating AOPs with bioprocesses. Therefore, AOPs are applicable as a pre-treatment stage prior to biodegradation, from both environmental and economic perspectives [101, 102]. The classical Fenton process relies on a redox reaction, in which ferrous ion (Fe^{2+}) and hydrogen peroxide (H_2O_2) react in an acidic medium to generate OH radicals. Fenton based processes are approved as reliable techniques that provide a high efficiency of eliminating target compounds from water and wastewater matrices and mineralizing pollutant molecules, which can be evaluated by observing a decrease in TOC or DOC (dissolved organic carbon). Regardless of the significant activity of AOPs, the degradation products of these processes are still toxic. Due to the differences in the operational parameters of Fenton based processes, their performance cannot be compared. Finally, Fenton based techniques can be affected by several operational parameters including pH and concentrations of Fe^{2+} and H_2O_2 [103].

Adsorption

Adsorption can be described as the attachment of atoms, ions, biomolecules, or molecules from a gas, liquid, or dissolved solid to an adsorbent. This process generates a thick layer (film) of the adsorbate on the surface of the adsorbent. Adsorption processes can be classified as ionic, physical, and chemical adsorption methods. Recently, a combination of nanotechnology and activated carbon for adsorption of pollutants has emerged [104]. In general, the mechanism of adsorption and the associated reactions can be unraveled by investigating the isotherms and kinetics of adsorption.

Adsorption method is better than degradation methods as it is safer, economic, and eco-friendly and does not produce toxic secondary products. Furthermore, adsorption processes are easy to use, require moderate operational conditions, and are economically affordable in operational view. Nevertheless, production of activated carbon is expensive and difficult [104, 105]. Activated carbon is one of the typical adsorbents for removal of pollutants from aqueous environments. With respect to the extent of pollution, the powdered or granulated form of activated carbon can be adopted. Activated carbon consists of about 70% to 90% carbon. The porous structure of this carbon presents a great surface area for adsorption of pollutant compounds. The unique properties of activated carbon originate from its high specific surface area, its mesoporous structure, and vast dispersion of its active sites. Despite the noticeable features of activated carbon, reducing carbon and its disposal to the environment demands some special considerations [106]. In addition to its high capacity of adsorbing organic compounds (such as antibiotics) from water and wastewater, activated carbon is highly efficient for eliminating inorganic compounds such as nitrate [107]. Activated carbon can be produced from agricultural wastes and applied to pollutant removal. The controlling factors of this process include the optimum dosage of the agricultural waste that would be converted to activated carbon, optimal adsorption conditions, and the types of materials that can be used for adsorption [108]. The ash used as an adsorbent can be produced from a wide spectrum of carbonaceous materials, such as wood, charcoal, walnut shell, pyrene, and agricultural wastes. Due to the high cost of other water treatment methods, some other options, *e.g.*, application of agricultural wastes and pyrene as adsorbents, have been considered as economic, efficient, and novel methods of pollutant removal from water and wastewater [109, 110].

Adsorption with Natural Materials

Adsorption is a physicochemical process in which a fluid is adsorbed on the surface of a solid absorbent. The fluid absorbed in this process is called adsorbate. Generally, the use of expensive adsorbents in water treatment and nitrate removal from water is not economically justified. Therefore, the use of low-cost natural adsorbents is increasing [111].

Activated carbon is known as a good absorbent in the removal of nitrate because of its high porosity and high specific surface area. However, it is not used in large scales due to its high cost [112]. Some studies have shown that agricultural wastes such as bagasse, corncob, and coconut shell provide a good source of activated carbon production. Biochar is a porous charcoal that is produced by thermal carbonization of the biomass. The high charge and high specific surface area makes this method a great use in water treatment [111].

Natural zeolites were first discovered by scientists in South Africa. Mike *et al.* showed that surfactant-modified zeolite has a high efficiency in removing nitrate from water. Surfactant-modified zeolites can absorb nitrates and negatively charged pollutants because of their positive charge surfaces [113]. An 80% removal efficiency of nitrate was achieved using acid-activated bentonite clay. Modification of adsorbent with acid improves absorbent capacity and increases the active sites and the specific surface area [114]. Physical, chemical, and biological treatments are three methods used in adsorption of nitrate and nitrite from water and wastewater. AOPs often break down pollutants into simpler molecules or can lead to mineralization; however, these methods are very complex and costly. Biological methods also have high efficiency in eliminating these types of pollutants, but due to the high volume of sludge production, sludge disposal, and high costs, they are not suitable. Accordingly, the use of more efficient and economical methods is recommended. Physical techniques are among the most effective methods for treatment of this wastewater [115 - 117].

In several studies, bio-adsorbents have been used to remove nitrogen from water and wastewater [118 - 120]. For example, Biochar is a stable material that is resistant to mineralization and is also rich in carbon, which is considered to be an advantage in the process of adsorption [121].

CONCLUSION

Different methods have been used to remove nitrates, each having their own advantages and disadvantages. In this regard, biological denitrification is a good method with many advantages. However, its long initial startup period and the possibility of water contamination with microbial substances make it inadequate for drinking water treatment; thus, it is more often used to remove nitrate from sewage. Chemical denitrification has no high practical application due to the low efficiency, high cost of chemicals, and the long reaction time. Water purification with ED is used in special cases due to membrane fouling, the need for pretreatment process, and the cost of energy required for the process. Among the treatment processes, ion exchange and RO are widely used for removing nitrates from drinking water. A short startup period, requiring low temperatures, and high efficiency, especially for low-volume systems, are advantages of these processes. The RO process requires pretreatment, is further used for desalination, and is less used for nitrate removal from water. The removal of nitrate from drinking water by the ion exchange method seems to be more acceptable, but biological methods are more convenient for wastewater treatment.

CONSENT FOR PUBLICATION

Not applicable.

REFERENCES

[1]　Shrimali M, Singh KP. New methods of nitrate removal from water. Environ Pollut 2001; 112(3): 351-9.
[http://dx.doi.org/10.1016/S0269-7491(00)00147-0] [PMID: 11291441]

[2]　Cao S. [Studies on the reactivity activation of zero-valent iron (ZVI) with hydrogen peroxide for nitrate reduction in mine water]. Master thesis, LUT-University, 2016; 18-21. http://urn.fi/URN:NBN:fi- fe2016090123374

[3]　Patil I, Husain M, Rahane V. Ground water nitrate removal by using 'Chitosan'as an adsorbent. Intern J Mod Eng Res 2013; 1(3): 346-9.

[4]　Sumino T, Isaka K, Ikuta H, Saiki Y, Yokota T. Nitrogen removal from wastewater using simultaneous nitrate reduction and anaerobic ammonium oxidation in single reactor. J Biosci Bioeng 2006; 102(4): 346-51.
[http://dx.doi.org/10.1263/jbb.102.346] [PMID: 17116583]

[5]　Hudak P. Regional trends in nitrate content of Texas groundwater. J Hydrol (Amst) 2000; 228(1-2): 37-47.
[http://dx.doi.org/10.1016/S0022-1694(99)00206-1]

[6]　Ransom KM. Bayesian and Machine Learning Methods for the Analysis of Nitrate in Groundwater in the Central Valley, California, USA. Davis: University of California 2017.

[7]　Fataei E, Sharifi AS, Kourandeh HHP, Seyyed A, Sharifi STSS. Nitrate removal from drinking water in laboratory-scale using iron and sand nanoparticles. Int J Biosci 2013; 3(10): 256-61.
[http://dx.doi.org/10.12692/ijb/3.10.256-261]

[8] Ghafari S, Hasan M, Aroua MK. Bio-electrochemical removal of nitrate from water and wastewater--a review. Bioresour Technol 2008; 99(10): 3965-74.
[http://dx.doi.org/10.1016/j.biortech.2007.05.026] [PMID: 17600700]

[9] Foglar L, Briški F, Sipos L, Vuković M. High nitrate removal from synthetic wastewater with the mixed bacterial culture. Bioresour Technol 2005; 96(8): 879-88.
[http://dx.doi.org/10.1016/j.biortech.2004.09.001] [PMID: 15627558]

[10] Inoue-Choi M, Jones RR, Anderson KE, *et al.* Nitrate and nitrite ingestion and risk of ovarian cancer among postmenopausal women in Iowa. Int J Cancer 2015; 137(1): 173-82.
[http://dx.doi.org/10.1002/ijc.29365] [PMID: 25430487]

[11] Jones RR, Weyer PJ, Dellavalle CT, *et al.* Nitrate from drinking water and diet and bladder cancer among postmenopausal women in Iowa. Environ Health Pers (EHP) 2016; 124(11): 1751.
[http://dx.doi.org/10.1289/EHP191]

[12] Kapoor A, Viraraghavan T. Nitrate removal from drinking water. J Environ Eng 1997; 123(4): 371-80.
[http://dx.doi.org/10.1061/(ASCE)0733-9372(1997)123:4(371)]

[13] Till BA, Weathers LJ, Alvarez PJ. Fe (0)-supported autotrophic denitrification. Environ Sci Technol 1998; 32(5): 634-9.
[http://dx.doi.org/10.1021/es9707769]

[14] Bhatnagar A, Kumar E, Sillanpää M. Nitrate removal from water by nano-alumina: Characterization and sorption studies. Chem Eng J 2010; 163(3): 317-23.
[http://dx.doi.org/10.1016/j.cej.2010.08.008]

[15] Yaghmaeian K, Mehrafrooz A, Pouraslan A, Akbarpour M, Akbarpour B. Removal of Nitrate from Aqueous Solutions by Starch Stabilized nano Zero-Valent Iron (nZVI). J Environ Health Eng 2016; 3(4): 323-35.
[http://dx.doi.org/10.18869/acadpub.jehe.3.4.323]

[16] Kabay N, Yüksel M, Samatya S, Arar Ö, Yüksel Ü. Removal of nitrate from ground water by a hybrid process combining electrodialysis and ion exchange processes. Sep Sci Technol 2007; 42(12): 2615-27.
[http://dx.doi.org/10.1080/01496390701511374]

[17] Westerhoff P, Doudrick K. Nitrates in Groundwater: Treatment technologies for today and tomorrow. Southwest Hydrology 2009; 8: 30-1.

[18] Jensen V, Darby J, Seidel C, Gorman C. Technical Report 6 in: Addressing Nitrate in California's Drinking Water with a Focus on Tulare Lake Basin and Salinas Valley Groundwater. Report for the State Water Resources Control Board Report to the Legislature. Davis: Center for Watershed Sciences, University of California 2012.

[19] Jensen VB. Nitrate in potable water supplies: Treatment options, incidence of occurrence, and brine waste management. Davis: University of California 2015.

[20] Samatya S, Kabay N, Yüksel Ü, Arda M, Yüksel M. Removal of nitrate from aqueous solution by nitrate selective ion exchange resins. React Funct Polym 2006; 66(11): 1206-14.
[http://dx.doi.org/10.1016/j.reactfunctpolym.2006.03.009]

[21] Milmile SN, Pande JV, Karmakar S, Bansiwal A, Chakrabarti T, Biniwale RB. Equilibrium isotherm and kinetic modeling of the adsorption of nitrates by anion exchange Indion NSSR resin. Desalination 2011; 276(1-3): 38-44.
[http://dx.doi.org/10.1016/j.desal.2011.03.015]

[22] Capodaglio AG, Hlavínek P, Raboni M. Physico-chemical technologies for nitrogen removal from wastewaters: A review 2015. http://www.scielo.br/scielo.php?script=sci_arttext&pid=S1980-993 X2015000300481&lng=en

[23] Crittenden J, Trussell R, Hand D, Howe K, Tchobanoglous G. Water treatment, principles and design,

MWH. NY: John Wiley & Sons 2005; p. 1298.

[24]   Kemper JM, Westerhoff P, Dotson A, Mitch WA. Nitrosamine, dimethylnitramine, and chloropicrin formation during strong base anion-exchange treatment. Environ Sci Technol 2009; 43(2): 466-72. [http://dx.doi.org/10.1021/es802460n] [PMID: 19238981]

[25]   Dvorak BI, Skipton SO. Drinking water treatment: Reverse osmosis University of Nebraska–Lincoln Extension. Institute of Agriculture and Natural Resources 2008.

[26]   Darbi A, Viraraghavan T, Butler R, Corkal D. Pilot-scale evaluation of select nitrate removal technologies. J Environ Sci Health A Tox Hazard Subst Environ Eng 2003; 38(9): 1703-15. [http://dx.doi.org/10.1081/ESE-120022873] [PMID: 12940476]

[27]   Bebee J, Alspach B, Diamond S, Miller C, O'Neill T, Parker C, Eds. Using reverse osmosis and ion exchange as parallel processes to remove high nitrate levels. In: AWWA Annual Conference; 2003.

[28]   Bergman R, Bergman R. Reverse osmosis and nanofiltration. American Water Works Association 2007.

[29]   Duranceau S, Taylor J, Alexander A. Water Quality and Treatment Denver, Colorado: AWWA 2011; 11.4-9.

[30]   Sharma SK, Sobti RC. Nitrate removal from ground water: a review. J Chem-Ny 2012; 9(4): 1667-75.

[31]   Häyrynen K, Pongrácz E, Väisänen V, *et al.* Concentration of ammonium and nitrate from mine water by reverse osmosis and nanofiltration. Desalination 2009; 240(1-3): 280-9. [http://dx.doi.org/10.1016/j.desal.2008.02.027]

[32]   Richards LA, Vuachère M, Schäfer AI. Impact of pH on the removal of fluoride, nitrate and boron by nanofiltration/reverse osmosis. Desalination 2010; 261(3): 331-7. [http://dx.doi.org/10.1016/j.desal.2010.06.025]

[33]   Bi J, Peng C, Xu H, Ahmed A-S. Removal of nitrate from groundwater using the technology of electrodialysis and electrodeionization. Desalination Water Treat 2011; 34(1-3): 394-401. [http://dx.doi.org/10.5004/dwt.2011.2891]

[34]   Wilson JR. Demineralization by electrodialysis. Butterworths Scientific Publications 1960.

[35]   Banasiak LJ, Schäfer AI. Removal of boron, fluoride and nitrate by electrodialysis in the presence of organic matter. J Membr Sci 2009; 334(1-2): 101-9. [http://dx.doi.org/10.1016/j.memsci.2009.02.020]

[36]   Prato T, Parent RG, Eds. Nitrate and nitrite removal from municipal drinking water supplies with electrodialysis reversal Proceedings of 1993 AWWA Membrane Conference: USA: Baltimore, Maryland, 1993.

[37]   Elmidaoui A, Elhannouni F, Sahli MM, *et al.* Pollution of nitrate in Moroccan ground water: removal by electrodialysis. Desalination 2001; 136(1-3): 325-32. [http://dx.doi.org/10.1016/S0011-9164(01)00195-3]

[38]   Abou-Shady A, Peng C, Bi J, Xu H. Recovery of Pb (II) and removal of NO_3^- from aqueous solutions using integrated electrodialysis, electrolysis, and adsorption process. Desalination 2012; 286: 304-15. [http://dx.doi.org/10.1016/j.desal.2011.11.041]

[39]   Strathmann H. Electrodialysis, a mature technology with a multitude of new applications. Desalination 2010; 264(3): 268-88. [http://dx.doi.org/10.1016/j.desal.2010.04.069]

[40]   Tchobanoglous G, Burton FL, Stensel HD. Wastewater Engineering: Treatment and Reuse (4th edn, revised).New York: Metcalf & Eddy Inc. and McGraw Hill Companies Inc. 2003.

[41]   Orentlicher M. Overview of nitrogen removal technologies and application/use of associated end products.Proceedings of Got Manure Enhancing Environmental and Economic Sustainability Conference. , New York, NY, USA 2012; pp. 148-58. [Accessed on 1 May 2012];

http://www.ansci.cornell.edu/prodairy/gotmanure/index.html

[42] Zhang T, Ding L, Ren H. Pretreatment of ammonium removal from landfill leachate by chemical precipitation. J Hazard Mater 2009; 166(2-3): 911-5.
[http://dx.doi.org/10.1016/j.jhazmat.2008.11.101] [PMID: 19135791]

[43] Asada T, Ohkubo T, Kawata K, Oikawa K. Ammonia adsorption on bamboo charcoal with acid treatment. J Health Sci 2006; 52(5): 585-9.
[http://dx.doi.org/10.1248/jhs.52.585]

[44] Tchobanoglous G. Fundamentals of biological treatment. Wastewater engineering: Treatment and reuse 2003; 611-35. https://ci.nii.ac.jp/naid/20001349940/en/

[45] Parastar S, Nasseri S, Borji SH, *et al.* Application of Ag-doped TiO_2 nanoparticle prepared by photodeposition method for nitrate photocatalytic removal from aqueous solutions. Desalination Water Treat 2013; 51(37-39): 7137-44.
[http://dx.doi.org/10.1080/19443994.2013.771288]

[46] Bui THL. [Removal of nitrate from water and wastewater by ammonium-functionalized SBA-16 mesoporous silica]. Master thesis, Laval University; 2013: 1-26. Available from: https://corpus.ulaval.ca/jspui/handle/20.500.11794/23966?locale=en

[47] Mohammadian FZ, Rabieh S, Zavar MH, Bagheri M. Preparation of Fe/activated carbon directly from orange peel and its application in removal of nitrate from aqueous solutions. J App Chem 2018; 12(45): 41-9.

[48] Savage N, Diallo MS. Nanomaterials and water purification: opportunities and challenges. J Nanopart Res 2005; 7(4-5): 331-42.
[http://dx.doi.org/10.1007/s11051-005-7523-5]

[49] Van der Bruggen B, Vandecasteele C. Removal of pollutants from surface water and groundwater by nanofiltration: overview of possible applications in the drinking water industry. Environ Pollut 2003; 122(3): 435-45.
[http://dx.doi.org/10.1016/S0269-7491(02)00308-1] [PMID: 12547533]

[50] Fu F, Dionysiou DD, Liu H. The use of zero-valent iron for groundwater remediation and wastewater treatment: a review. J Hazard Mater 2014; 267: 194-205.
[http://dx.doi.org/10.1016/j.jhazmat.2013.12.062] [PMID: 24457611]

[51] Sun Y-P, Li X-Q, Zhang W-X, Wang HP. A method for the preparation of stable dispersion of zero-valent iron nanoparticles. Colloids Surf A Physicochem Eng Asp 2007; 308(1-3): 60-6.
[http://dx.doi.org/10.1016/j.colsurfa.2007.05.029]

[52] Tiraferri A, Chen KL, Sethi R, Elimelech M. Reduced aggregation and sedimentation of zero-valent iron nanoparticles in the presence of guar gum. J Colloid Interface Sci 2008; 324(1-2): 71-9.
[http://dx.doi.org/10.1016/j.jcis.2008.04.064] [PMID: 18508073]

[53] Jiemvarangkul P, Zhang W-x, Lien H-L. Enhanced transport of polyelectrolyte stabilized nanoscale zero-valent iron (nZVI) in porous media. Chem Eng J 2011; 170(2-3): 482-91.
[http://dx.doi.org/10.1016/j.cej.2011.02.065]

[54] Lin N, Huang J, Dufresne A. Preparation, properties and applications of polysaccharide nanocrystals in advanced functional nanomaterials: a review. Nanoscale 2012; 4(11): 3274-94.
[http://dx.doi.org/10.1039/c2nr30260h] [PMID: 22565323]

[55] Sargent JF. Nanotechnology and Environmental Health and Safety: Issues for Consideration. DIANE Publishing 2008.

[56] Xu D, Li Y, Yin L, Ji Y, Niu J, Yu Y. Electrochemical removal of nitrate in industrial wastewater. Front Environ Sci Eng 2018; 12(1): 9.
[http://dx.doi.org/10.1007/s11783-018-1033-z]

[57] Mook W, Chakrabarti M, Aroua M, *et al.* Removal of total ammonia nitrogen (TAN), nitrate and total

organic carbon (TOC) from aquaculture wastewater using electrochemical technology: A review. Desalination 2012; 285: 1-13.
[http://dx.doi.org/10.1016/j.desal.2011.09.029]

[58]   Govindan K, Noel M, Mohan R. Removal of nitrate ion from water by electrochemical approaches. J Water Process Eng 2015; 6: 58-63.
[http://dx.doi.org/10.1016/j.jwpe.2015.02.008]

[59]   Jia Y-H, Tran H-T, Kim D-H, *et al.* Simultaneous organics removal and bio-electrochemical denitrification in microbial fuel cells. Bioprocess Biosyst Eng 2008; 31(4): 315-21.
[http://dx.doi.org/10.1007/s00449-007-0164-6] [PMID: 17909860]

[60]   Jara C, Martínez-Huitle C, Torres-Palma R. Distribution of nitrogen ions generated in the electrochemical oxidation of nitrogen containing organic compounds. Port Electrochem Acta 2009; 27(3): 203-13.
[http://dx.doi.org/10.4152/pea.200903203]

[61]   Pletcher D. Guide to electrochemical technology for synthesis, separation and pollution control. Electrosynthesis Company Inc. 1999.

[62]   Feng C, Sugiura N, Shimada S, Maekawa T. Development of a high performance electrochemical wastewater treatment system. J Hazard Mater 2003; 103(1-2): 65-78.
[http://dx.doi.org/10.1016/S0304-3894(03)00222-X] [PMID: 14568697]

[63]   Walsh F. Electrochemical technology for environmental treatment and clean energy conversion. Pure Appl Chem 2001; 73(12): 1819-37.
[http://dx.doi.org/10.1351/pac200173121819]

[64]   Wang G, Zhang L, Zhang J. A review of electrode materials for electrochemical supercapacitors. Chem Soc Rev 2012; 41(2): 797-828.
[http://dx.doi.org/10.1039/C1CS15060J] [PMID: 21779609]

[65]   Su L, Li K, Zhang H, *et al.* Electrochemical nitrate reduction by using a novel Co_3O_4/Ti cathode. Water Res 2017; 120: 1-11.
[http://dx.doi.org/10.1016/j.watres.2017.04.069] [PMID: 28478288]

[66]   Anglada A, Urtiaga A, Ortiz I. Contributions of electrochemical oxidation to waste-water treatment: fundamentals and review of applications. J Chem Technol Biotechnol 2009; 84(12): 1747-55.
[http://dx.doi.org/10.1002/jctb.2214]

[67]   Matsunaga A, Yasuhara A. Dechlorination of polychlorinated organic compounds by electrochemical reduction with naphthalene radical anion as mediator. Chemosphere 2005; 59(10): 1487-96.
[http://dx.doi.org/10.1016/j.chemosphere.2004.12.045] [PMID: 15876391]

[68]   Jiang C, Liu L, Crittenden JC. An electrochemical process that uses an $Fe0/TiO_2$ cathode to degrade typical dyes and antibiotics and a bio-anode that produces electricity. Front Environ Sci Eng 2016; 10(4): 15.
[http://dx.doi.org/10.1007/s11783-016-0860-z]

[69]   Ghazouani M, Akrout H, Bousselmi L. Efficiency of electrochemical denitrification using electrolysis cell containing BDD electrode. Desalination Water Treat 2015; 53(4): 1107-17.
[http://dx.doi.org/10.1080/19443994.2014.884473]

[70]   Georgeaud V, Diamand A, Borrut D, Grange D, Coste M. Electrochemical treatment of wastewater polluted by nitrate: selective reduction to N_2 on boron-doped diamond cathode. Water Sci Technol 2011; 63(2): 206-12.
[http://dx.doi.org/10.2166/wst.2011.034] [PMID: 21252421]

[71]   Ghazouani M, Akrout H, Bousselmi L. Nitrate and carbon matter removals from real effluents using Si/BDD electrode. Environ Sci Pollut Res Int 2017; 24(11): 9895-906.
[http://dx.doi.org/10.1007/s11356-016-7563-7] [PMID: 27623854]

[72]   Fabiańska A, Białk-Bielińska A, Stepnowski P, Stolte S, Siedlecka EM. Electrochemical degradation

of sulfonamides at BDD electrode: kinetics, reaction pathway and eco-toxicity evaluation. J Hazard Mater 2014; 280: 579-87.
[http://dx.doi.org/10.1016/j.jhazmat.2014.08.050] [PMID: 25215656]

[73]   Su JF, Ruzybayev I, Shah I, Huang C. The electrochemical reduction of nitrate over micro-architectured metal electrodes with stainless steel scaffold. Appl Catal B 2016; 180: 199-209.
[http://dx.doi.org/10.1016/j.apcatb.2015.06.028]

[74]   Dogonadze R, Ulstrup J, Kharkats YI. A theory of electrode reactions through bridge transition states; bridges with a discrete electronic spectrum. Electroanal Chem 1972; 39(1): 47-61.
[http://dx.doi.org/10.1016/S0022-0728(72)80475-3]

[75]   Szpyrkowicz L, Daniele S, Radaelli M, Specchia S. Removal of NO_3^- from water by electrochemical reduction in different reactor configurations. Appl Catal B 2006; 66(1-2): 40-50.
[http://dx.doi.org/10.1016/j.apcatb.2006.02.020]

[76]   Hasnat M, Rashed M, Aoun SB, *et al.* Dissimilar catalytic trails of nitrate reduction on Cu-modified Pt surface immobilized on H^+ conducting solid polymer. J Mol Catal Chem 2014; 383: 243-8.
[http://dx.doi.org/10.1016/j.molcata.2013.12.015]

[77]   Zaroual Z, Azzi M, Saib N, Chaînet E. Contribution to the study of electrocoagulation mechanism in basic textile effluent. J Hazard Mater 2006; 131(1-3): 73-8.
[http://dx.doi.org/10.1016/j.jhazmat.2005.09.021] [PMID: 16243434]

[78]   Bazrafshan E, Kord Mostafapour F, Farzadkia M, Ownagh K, Jaafari Mansurian H. Application of Combined Chemical Coagulation-Electro Coagulation Process for Treatment of the Zahedan Cattle Slaughterhouse Wastewater. IJHE 2012; 5(3): 283-94.

[79]   Emamjomeh M, Sivakumar M. Electrocoagulation (EC) technology for nitrate removal. In: Khanna N, Ed. Environmental Postgrad Conference; Environmental change: Making it Happen 2005; 1-8.
http://www.uow.edu.au/1678

[80]   AhmadiMoghadam M, Amiri H. Investigation of TOC Removal from IndustrialWastewaters using Electrocoagulation Process. Iran J Health & Environ 2010; 3(2): 185-94.

[81]   Razavi M, Saeedi M, Jabaari E. Comparison of the cost and efficiency of Aluminum and Iron electrodes application in the removal of phosphate, nitrate, and COD from laundry wastewater using electrocoagulation process. Iran J Health & Environ 2013; 6(3): 265-76.

[82]   Habib MA, Moghaddam SMRA, Arami M, Hashemi SH. Optimization of the electrocoagulation process for removal of Cr (VI) using Taguchi method. Ab va Fazilab 2012; 22-4.

[83]   Mithra SS, Ramesh ST, Gandhimathi R, Nidheesh PV. Studies on the removal of phosphate from water by electrocoagulation with aluminium plate electrodes. Environ Eng Manag J 2017; 16(10)

[84]   İrdemez Ş, Demircioğlu N, Yıldız YŞ, Bingül Z. The effects of current density and phosphate concentration on phosphate removal from wastewater by electrocoagulation using aluminum and iron plate electrodes. Separ Purif Tech 2006; 52(2): 218-23.
[http://dx.doi.org/10.1016/j.seppur.2006.04.008]

[85]   Koparal AS, Oğütveren ÜB. Removal of nitrate from water by electroreduction and electrocoagulation. J Hazard Mater 2002; 89(1): 83-94.
[http://dx.doi.org/10.1016/S0304-3894(01)00301-6] [PMID: 11734348]

[86]   Emamjomeh M, Sivakumar M, Schafer A. Fluoride removal by using a batch electrocoagulation reactor. In: Seventh Annual Environmental Engineering Research Event (EERE) Conference; 1st–4th December; Marysville, Victoria, Australia. 2003; pp. 143-52.http://www.uow.edu.au/

[87]   Emamjomeh MM, Sivakumar M. Effects of calcium ion on enhanced defluoridation by Electrocoagulation/flotation (ECF) process. In: Eighth Annual Environmental Engineering Research Event (EERE) Conference; 6th–9th December; Wollongong, New South Wales, Australia. 2004; pp. 263-74.http://www.uow.edu.au/

[88]   Mollah MYA, Schennach R, Parga JR, Cocke DL. Electrocoagulation (EC)--science and applications. J Hazard Mater 2001; 84(1): 29-41.
[http://dx.doi.org/10.1016/S0304-3894(01)00176-5] [PMID: 11376882]

[89]   Bennajah M, Gourich B, Essadki AH, Vial C, Delmas H. Defluoridation of Morocco drinking water by electrocoagulation/electroflottation in an electrochemical external-loop airlift reactor. Chem Eng J 2009; 148(1): 122-31.
[http://dx.doi.org/10.1016/j.cej.2008.08.014]

[90]   Emamjomeh MM, Sivakumar M. Review of pollutants removed by electrocoagulation and electrocoagulation/flotation processes. J Environ Manage 2009; 90(5): 1663-79.
[http://dx.doi.org/10.1016/j.jenvman.2008.12.011] [PMID: 19181438]

[91]   Kumar NS, Goel S. Factors influencing arsenic and nitrate removal from drinking water in a continuous flow electrocoagulation (EC) process. J Hazard Mater 2010; 173(1-3): 528-33.
[http://dx.doi.org/10.1016/j.jhazmat.2009.08.117] [PMID: 19766389]

[92]   Majlesi M, Mohseny SM, Sardar M, Golmohammadi S, Sheikhmohammadi A. Improvement of aqueous nitrate removal by using continuous electrocoagulation/electroflotation unit with vertical monopolar electrodes. Sustain Environ Res 2016; 26(6): 287-90.
[http://dx.doi.org/10.1016/j.serj.2016.09.002]

[93]   Lacasa E, Cañizares P, Sáez C, Fernández FJ, Rodrigo MA. Removal of nitrates from groundwater by electrocoagulation. Chem Eng J 2011; 171(3): 1012-7.
[http://dx.doi.org/10.1016/j.cej.2011.04.053]

[94]   Parsons S. Advanced oxidation processes for water and wastewater treatment. IWA Publishing 2004.

[95]   Hapeshi E, Fotiou I, Fatta-Kassinos D. Sonophotocatalytic treatment of ofloxacin in secondary treated effluent and elucidation of its transformation products. Chem Eng J 2013; 224: 96-105.
[http://dx.doi.org/10.1016/j.cej.2012.11.048]

[96]   Barrabés N, Sá J. Catalytic nitrate removal from water, past, present and future perspectives. Appl Catal B 2011; 104(1-2): 1-5.
[http://dx.doi.org/10.1016/j.apcatb.2011.03.011]

[97]   Skoumal M, Cabot P-L, Centellas F, et al. Mineralization of paracetamol by ozonation catalyzed with Fe^{2+}, Cu^{2+} and UVA light. Appl Catal B 2006; 66(3-4): 228-40.
[http://dx.doi.org/10.1016/j.apcatb.2006.03.016]

[98]   Glaze WH, Kang J-W, Chapin DH. The chemistry of water treatment processes involving ozone, hydrogen peroxide and ultraviolet radiation. Taylor and Francis 1987.
[http://dx.doi.org/10.1080/01919518708552148]

[99]   Kritzer P, Dinjus E. An assessment of supercritical water oxidation (SCWO): existing problems, possible solutions and new reactor concepts. Chem Eng J 2001; 83(3): 207-14.
[http://dx.doi.org/10.1016/S1385-8947(00)00255-2]

[100] Karimi B, Rajaei MS. Evaluating of Nitrate removal by adsorption/Fe/H_2O_2 process from water: Kinetics and operation parameters. Feyz J Kashan Uni Med Sci 2013; 16.

[101] Cañizares P, Paz R, Sáez C, Rodrigo MA. Costs of the electrochemical oxidation of wastewaters: a comparison with ozonation and Fenton oxidation processes. J Environ Manage 2009; 90(1): 410-20.
[http://dx.doi.org/10.1016/j.jenvman.2007.10.010] [PMID: 18082930]

[102] Poyatos JM, Muñio M, Almecija M, Torres J, Hontoria E, Osorio F. Advanced oxidation processes for wastewater treatment: state of the art. Water Air Soil Pollut 2010; 205(1-4): 187.
[http://dx.doi.org/10.1007/s11270-009-0065-1]

[103] Brillas E, Sirés I, Oturan MA. Electro-Fenton process and related electrochemical technologies based on Fenton's reaction chemistry. Chem Rev 2009; 109(12): 6570-631.
[http://dx.doi.org/10.1021/cr900136g] [PMID: 19839579]

[104] Kakavandi B, Esrafili A, Mohseni-Bandpi A, Jonidi Jafari A, Rezaei Kalantary R. Magnetic $Fe_3O_4@C$ nanoparticles as adsorbents for removal of amoxicillin from aqueous solution. Water Sci Technol 2014; 69(1): 147-55.
[http://dx.doi.org/10.2166/wst.2013.568] [PMID: 24434981]

[105] Moussavi G, Alahabadi A, Yaghmaeian K, Eskandari M. Preparation, characterization and adsorption potential of the NH_4 Cl-induced activated carbon for the removal of amoxicillin antibiotic from water. Chem Eng J 2013; 217: 119-28.
[http://dx.doi.org/10.1016/j.cej.2012.11.069]

[106] Snyder S, Adham S, Redding A, *et al.* Role of membranes and activated carbon in the removal of endocrine disruptors and pharmaceuticals. Desalination 2007; 202(1-3): 156-81.
[http://dx.doi.org/10.1016/j.desal.2005.12.052]

[107] Shan D, Deng S, Zhao T, *et al.* Preparation of ultrafine magnetic biochar and activated carbon for pharmaceutical adsorption and subsequent degradation by ball milling. J Hazard Mater 2016; 305: 156-63.
[http://dx.doi.org/10.1016/j.jhazmat.2015.11.047] [PMID: 26685062]

[108] Zhou Y, Zhang L, Cheng Z. Removal of organic pollutants from aqueous solution using agricultural wastes: A review. J Mol Liq 2015; 212: 739-62.
[http://dx.doi.org/10.1016/j.molliq.2015.10.023]

[109] Mahvi AH, Maleki A. The use of agricultural waste for removal of phenol in water environments. HM J 2006; 10: 85.

[110] Jafari K, Heidari M, Rahmanian O. Wastewater treatment for Amoxicillin removal using magnetic adsorbent synthesized by ultrasound process. Ultrason Sonochem 2018; 45: 248-56.
[http://dx.doi.org/10.1016/j.ultsonch.2018.03.018] [PMID: 29705319]

[111] Mishra PC, Patel RK. Use of agricultural waste for the removal of nitrate-nitrogen from aqueous medium. J Environ Manage 2009; 90(1): 519-22.
[http://dx.doi.org/10.1016/j.jenvman.2007.12.003] [PMID: 18294755]

[112] Liang MN, Zeng HH, Zhu YN, Xu ZL, Liu HL, Eds. Adsorption removal of phosphorus from aqueous solution by the activated carbon prepared from sugarcane bagasse.Adv Mater Res-Switz, Trans Tech Publications Ltd. 2011; 183-185: pp. 1046-50.
[http://dx.doi.org/10.4028/www.scientific.net/AMR.183-185.1046]

[113] Masukume M, Onyango MS, Aoyi O, Otieno F. Nitrate removal from groundwater using modified natural zeolite. Water SA 2010; 36(5): 655-2. Available at: http://www.ewisa. co.za/literature/files/144 97%20Masukume.pdf
[http://dx.doi.org/10.4314/wsa.v36i5.61999]

[114] Wasse Bekele GF, Fernandez N. Removal of nitrate ion from aqueous solution by modified Ethiopian bentonite clay. IJRPC 2014; 4(1): 192-201.

[115] Lin S-H, Juang R-S. Adsorption of phenol and its derivatives from water using synthetic resins and low-cost natural adsorbents: a review. J Environ Manage 2009; 90(3): 1336-49.
[http://dx.doi.org/10.1016/j.jenvman.2008.09.003] [PMID: 18995949]

[116] Homem V, Santos L. Degradation and removal methods of antibiotics from aqueous matrices--a review. J Environ Manage 2011; 92(10): 2304-47.
[http://dx.doi.org/10.1016/j.jenvman.2011.05.023] [PMID: 21680081]

[117] Bhatnagar A, Sillanpää M. A review of emerging adsorbents for nitrate removal from water. Chem Eng J 2011; 168(2): 493-504.
[http://dx.doi.org/10.1016/j.cej.2011.01.103]

[118] Zhang M, Gao B, Yao Y, Xue Y, Inyang M. Synthesis of porous MgO-biochar nanocomposites for removal of phosphate and nitrate from aqueous solutions. Chem Eng J 2012; 210: 26-32.
[http://dx.doi.org/10.1016/j.cej.2012.08.052]

[119]   Mizuta K, Matsumoto T, Hatate Y, Nishihara K, Nakanishi T. Removal of nitrate-nitrogen from drinking water using bamboo powder charcoal. Bioresour Technol 2004; 95(3): 255-7.
[http://dx.doi.org/10.1016/j.biortech.2004.02.015] [PMID: 15288267]

[120]   Song W, Gao B, Xu X, *et al.* Adsorption of nitrate from aqueous solution by magnetic amine-crosslinked biopolymer based corn stalk and its chemical regeneration property. J Hazard Mater 2016; 304: 280-90.
[http://dx.doi.org/10.1016/j.jhazmat.2015.10.073] [PMID: 26561752]

[121]   Wardle DA, Nilsson M-C, Zackrisson O. Fire-derived charcoal causes loss of forest humus. Science 2008; 320(5876): 629.
[http://dx.doi.org/10.1126/science.1154960] [PMID: 18451294]

CHAPTER 2

Biological Methods for Nitrogen Removal

Saeb Ahmadi[1,*], Edris Hosseinzade[2], Ashkan Sami Vand[3] and **Hooshyar Hossini[4]**

[1] *Department of Chemical Engineering, Tarbiat Modares University, Tehran, Iran*

[2] *Environmental and Occupational Hazards Control Research Center, Shahid Beheshti University of Medical Sciences, Tehran, Iran*

[3] *Department of Civil & Environmental Engineering, Shiraz University, Shiraz, Iran*

[4] *Department of Environmental Health Engineering, Faculty of Health, Kermanshah University of Medical Sciences, Kermanshah, Iran*

Abstract: Pollution of water resources, such as nitrogen compounds, nitrate, and ammonium is increasingly expanding. A wide range of physico-chemical and biological methods are conducted for decreasing nitrate compounds and the amount of wastewaters. Due to low selectivity of nitrate, high costs, and the need for pre and post purification processes, the physico-chemical method is not effective for this purpose. However, biological methods are highly efficient in the removal of nitrogen compounds, and due to their advantages, they are widely used in wastewater purification. These methods include simultaneous nitrification and denitrification (SND), Anammox, and partial nitrification which are reviewed in the following chapter.

Keywords: Anammox, Denitrification, Nitrification, Partial Nitrification, SND.

INTRODUCTION

Biological processes have high efficiency and low costs. So, they are widely used in the removal of nitrate and nitrogen compounds from water. Physico-chemical methods have low efficiency nitrate removal from wastewaters due to their low selectivity of nitrate, operational restrictions, more concentrated wastewater generation, the need for pre and post treatment processes, and high costs. The most important advantage of biological denitrification is substitution of nitrate with harmless nitrogen gas. Based on the type of electron donor, this method is categorized into two major types of heterotrophic and autotrophic. In this chapter,

* **Corresponding author Saeb Ahmadi:** Department of Chemical Engineering, Tarbiat Modares University, Tehran, Iran; Tel: +982182883979; E-mail: saeb.ahmadi@gmail.com

Edris Hoseinzadeh (Ed.)

different kinds of biological methods used for the removal of nitrogen compounds, especially nitrate and ammonia, are analyzed. Also, their advantages and disadvantages are discussed.

Nitrification

Nitrification refers to the process of biological conversion of ammonium to nitrate. This process uses *Nitrosomonas* bacteria to convert ammonium and ammonia into nitrite in two steps. Eventually, the process of nitrite conversion to nitrate is done by the use of *Nitrobacter*. Nitrite concentration is mostly low due to high speed of reactions.

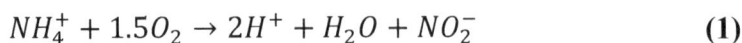

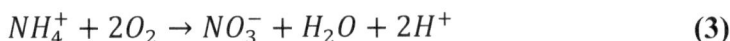

$$NH_4^+ + 1.5O_2 \rightarrow 2H^+ + H_2O + NO_2^- \tag{1}$$

$$NO_2^- + 0.5O_2 \rightarrow NO_3^- \tag{2}$$

$$NH_4^+ + 2O_2 \rightarrow NO_3^- + H_2O + 2H^+ \tag{3}$$

An efficient nitrification process requires a long retention time, a low food to microorganisms ratio, and sufficient amount of buffer [1].

Effective Factors at Nitrification Process

These factors are so variable including the presence of ammonia, dissolved oxygen (DO), pH, light, temperature, alkalinity, inorganic, and organic carbon source. All of these factors affect the nitrifying bacteria.

Presence of Ammonia

Ammonia, nitrite, and urea are the external energy resources for nitrifying bacteria. Therefore, they are necessary for the nitrification process [2].

Dissolved Oxygen (DO)

Nitrifying bacteria are aerobic. So nitrification needs more than 1 mg/L oxygen. For an effective nitrification, each mg of ammonia, which converts to nitrite, requires 4.33 mg oxygen. Ammonia oxidizers use nitrite or nitrate as an alternative electron acceptor in case of low amount of oxygen. Such kind of processes, like corrosion by creating an aerobic microbial environment, can use some amount of oxygen, and consequently, increase process efficiency and output [3].

Temperature

Temperature increases the rate of biological reactions by affecting the rates of enzymatic reactions and substrate diffusion on biofilm. Nitrifiers grow in a temperature range from 4°C to 45°C. The best temperature range for this process is 30°C to 35°C. In a temperature higher than 40°C and lower than 10°C, the rate of nitrification process reaches to zero [4, 5]. In a pure culture, the optimum temperatures for nitrosmonas and nitrobacter are 35°C and 35-42°C, respectively. The optimum temperature for activated sludge, consisting of AOB and NOB, is nearly 30°C. Researchers introduced variable optimum temperatures for nitrification process [6].

Light

Visible and ultraviolet light have limiting effects on nitrification process. Alleman and Preston concluded that growth of nitrifying bacteria in attached biofilm, in deeper layers as compared to less light, is higher. Alleman *et al.* stated that nitrifying bacteria can recover themselves after encountering light more than 4 to 6 hours [7].

pH

Nitrification process occurs at a wide pH range (4.6 to 11.2) [5]. Acid is produced in the nitrification process. This acid increases the required pH for nitrifying bacteria. The nitrification rate decreases at a pH lower than 6. Therefore, the optimum pH for this process is 7.5 to 8.5; pH affects the nitrification process due to changes in the concentration of free ammonia which use as a substrates for ammonia oxidation [8, 9].

Alkalinity and Inorganic Carbon

In the oxidation process of ammonia to nitrate, 14 mg $CaCO_3$ is consumed for every mg of NH_4^+-N which is so high. Most amount of alkali is used in ammonia oxidation for the neutralization process of hydrogen ions. Nitrifying bacteria are autotrophic and use inorganic carbon as their carbon source. More than 50-100 mg/L alkali is required for ensuring sufficient buffer amount [3].

Organic Carbon

Due to the toxic nature of organic substances and the competition between nitrifying and heterotrophic bacteria for the existent nutrients, organic carbon negatively affects the nitrification process [10, 11].

Micro Nutrient Limitation

Generally, microbial growth depends on available level of nutrients. The growth of aerobic bacteria can be described as follows:

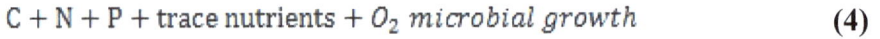

$$C + N + P + \text{trace nutrients} + O_2 \; microbial \; growth \qquad (4)$$

The nitrification process needs low quantities of such micronutrients as potassium, magnesium, calcium, and copper [12 - 14]. In addition, mineral carbon, phosphorus, and ammonia are considered to be macronutrients in nitrification process.

Biological Denitrification

Nitrate reduction to nitrogen gas in anoxic conditions consists of 4 steps, including: bacteria-induced reduction of NO_3^- into NO_2^-, then NO_2^- more reduced into NO, N_2O and N_2. In a suitable condition, this method reduces nitrate with high output and without any wastewater or the need for prolonged treatment. This method, described below, has a low cost [15, 16]:

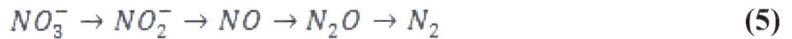

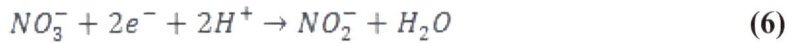

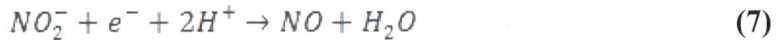

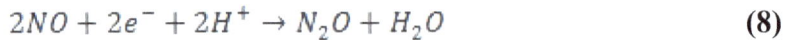

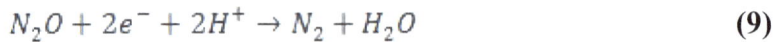

$$NO_3^- \rightarrow NO_2^- \rightarrow NO \rightarrow N_2O \rightarrow N_2 \qquad (5)$$

$$NO_3^- + 2e^- + 2H^+ \rightarrow NO_2^- + H_2O \qquad (6)$$

$$NO_2^- + e^- + 2H^+ \rightarrow NO + H_2O \qquad (7)$$

$$2NO + 2e^- + 2H^+ \rightarrow N_2O + H_2O \qquad (8)$$

$$N_2O + 2e^- + 2H^+ \rightarrow N_2 + H_2O \qquad (9)$$

Anoxic conditions are necessary for maximum efficiency of nitrate reduction in denitrification process. In case of higher than 0.1 mg/L oxygen concentration in the environment, denitrifying bacteria will reduce more oxygen rather than nitrate. Therefore, this will reduce the efficiency of denitrification process [16]. Heterotrophic denitrification uses organic carbon resources such as acetate, acetic acid, methanol and ethanol as its energy resources. The main advantage of this method is its high rate. On the other hand, its main disadvantages are considered to be the probability of microbial contamination and residual amount of carbon. This method can use industrial wastes, such as whey, as its energy resource. The reactions for methanol as a carbon resource are as follows [17]:

Step one (Nitrate to Nitrite): **(10)**

$$6NO_3^- + 3CH_3OH \rightarrow 6NO_2^- + 2CO_2 + 4H_2O \qquad (10)$$

Step two (Nitrite to Nitrogen gas): + **(11)**

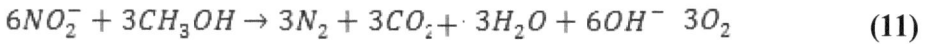

$$6NO_2^- + 3CH_3OH \rightarrow 3N_2 + 3CO_2 + 3H_2O + 6OH^- \ 3O_2 \tag{11}$$

Overall reaction:

$$6NO_3^- + 5CH_3OH + 3O_2 \rightarrow 3N_2 + 5CO_2 + 7H_2O + 6OH^- \tag{12}$$

This process has a lot of functions in biological treatment of wastewaters. Nitrate, as a terminal electron acceptor, is used by the bacteria. Therefore, it reduces to nitrogen gas. The efficiency of heterotrophic denitrification is higher than 95% [18]. Autotrophic bacteria, such as Thiomicrospira denitrificans and Thiobacillus denitrificans, can reduce nitrate to nitrogen gas. Microorganisms use hydrogen, iron, and sulfur as energy resources in autotrophic denitrification. They also use CO_2 and bicarbonate as the carbon resource. Balanced equations of reactions, in case of using sulfide, thiosulfate and Fe^{2+} as electron donors, are as follows [19 - 22]:

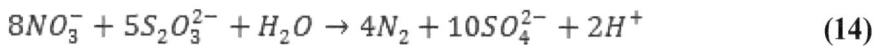

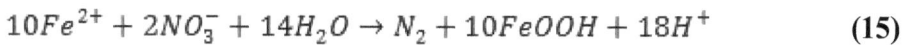

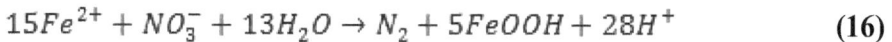

$$14NO_3^- + 5FeS_2 + 4H^+ \rightarrow 7N_2 + 10SO_4^{2-} + 5Fe^{2+} + 2H_2O \tag{13}$$

$$8NO_3^- + 5S_2O_3^{2-} + H_2O \rightarrow 4N_2 + 10SO_4^{2-} + 2H^+ \tag{14}$$

$$10Fe^{2+} + 2NO_3^- + 14H_2O \rightarrow N_2 + 10FeOOH + 18H^+ \tag{15}$$

$$15Fe^{2+} + NO_3^- + 13H_2O \rightarrow N_2 + 5FeOOH + 28H^+ \tag{16}$$

Using hydrogen as the electron donor in autotrophic denitrification has some advantages in comparison with other types of electrons: high selectivity for nitrate removal, less harmful by-product, low costs, possibility of its removal from water by air stripping process, and no need for prolonged treatments [23]. Its main disadvantage is possibility of creating a potentially explosive area due to the presence of hydrogen [24]. Stoichiometric equations of the autotrophic denitrifications with hydrogen are [25]:

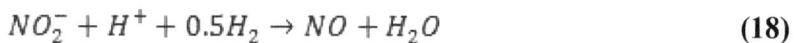

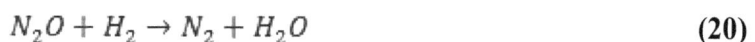

$$NO_3^- + H_2 \rightarrow NO_2^- + H_2O \tag{17}$$

$$NO_2^- + H^+ + 0.5H_2 \rightarrow NO + H_2O \tag{18}$$

$$2NO + H_2 \rightarrow N_2O + H_2O \tag{19}$$

$$N_2O + H_2 \rightarrow N_2 + H_2O \tag{20}$$

Overall reaction:

$$2NO_3^- + 5H_2 + 2H^+ \rightarrow N_2 + 6H_2O \tag{21}$$

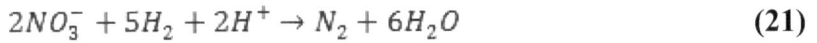

The advantages and disadvantages of the biological denitrification process are indicated in Table **1** [26].

Table 1. The advantages and disadvantages of the biological denitrification.

Advantages	Disadvantages
Less installation costs in comparison with other methods	Need for sufficient carbon resources
Easier transfer of the generated waste	Possibility of microbial infiltration for purifying the drinking water
High system stability	Long installation period
High efficiency in reducing nitrate level	Some factors such as: pH, DO, organic carbon and the presence of inhibitors in water which affect system function

The Effective Factors on Denitrification Process

Nitrate Concentration

Some researchers have investigated the effects of nitrate concentration on denitrification process and generated different results. Chang *et al.* stated that heterotrophic bacteria can remove high concentrations of nitrate [23]. Park *et al.* stated that the rate of nitrate removal will enhance with the increase of nitrate concentration in the range of 20 mg/L to 150 mg/L [27]. Zhou *et al.* found out that the total nitrate removal is possible in 10 mg/L concentration, but when concentration increase up to 30 mg/L, the removal process will have lower efficiency [28].

pH

Some researchers have announced the optimal pH for denitrification to be around 7.5 [29 - 31]. Denitrification is limited at the pH value above 7.5 due to nitrite accumulation. With a phosphate buffer, pH can remain around 7, but high concentration of phosphate buffer in biofilm will decrease the efficiency of denitrification process [32].

Temperature

Denitrification process requires 2°C to 50°C temperature range [33]. Researchers announced different optimal temperatures for this process; some of them announced this range to be 25°C to 35°C. Zhou *et al.* announced the temperature range of 25°C to 30°C as the optimum range for nitrate removal [28]. Rezania *et al.* observed an increase in denitrification rate with an increase in temperature from 12°C to 25°C. Furthermore, nitrate accumulation occurs in higher than 25°C temperature. The higher the temperature, the higher the efficiency of nitrate removal process [34].

Hardness and Alkalinity

Hardness and alkalinity affect the denitrification process negatively and reduce its rate. For example, Dries *et al.* [35] indicated that an increase in carbonate alkalinity (*i.e.*, $CaCO_3$) due to precipitation, will decrease the rate of denitrification process. Furthermore, precipitation decreases the biomass activity with restricting the mass transfer through biofilm [35].

Oxidation-Reduction Potential (ORP)

The optimal ORPs for denitrification process range from -50 mV to -250 mV [34]. In another study, the optimal ORP was announced to be -120 mV to -130 mV. Sakakibara *et al.* observed that an increase in HRT will increase ORP. Generally, with an increase in nitrate concentration, ORP increases as well, while an increase in hydrogen concentration decreases ORP [36].

Hydrogen Concentration

Chang *et al.* announced that if the concentration of dissolved hydrogen is below 0.2 mg/L, denitrification process will be incomplete [23]. Karanasios *et al.*, also, indicated that 0.4 mg/L to 0.8 mg/L hydrogen concentration will lead to nitrate removal [37]. Lee *et al.* reached 100% of nitrate removal with increasing hydrogen pressure from 0.45 atm to 0.56 atm [29]. A complete denitrification will occur in hydrogen pressure ranging from 0.2 atm to 0.55 atm [34].

Carbon Resource

Theoretically, the sufficient amount of carbon for a denitrification process is 0.2 mg C (in the form of bicarbonate) and 0.12-0.21 mg C (in form of CO_2) paramount to each mg of N [29]. In practice, more amount of carbon will be consumed. If the amount of C: N is higher than 0.21, nitrite accumulation occurs and a low C to N ratio will lead to incomplete denitrification [38, 39].

Reactor Technology

The attached growth systems are utilized in denitrification reactors due to lower cost of construction and operation and lower required space.

ANAMMOX

In Anammox process, ammonia converts into nitrogen gas using nitrite as an electron acceptor [40]. In 1992, Anammox bacteria were discovered in a denitrifying bioreactor [41]. Due to lower costs, the Anammox process is considered to be the main method for ammonia removal. Therefore, a wide range of researches have been carried out in this area. The balanced reactions for Anammox are [42]:

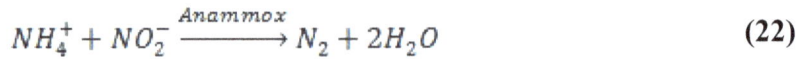

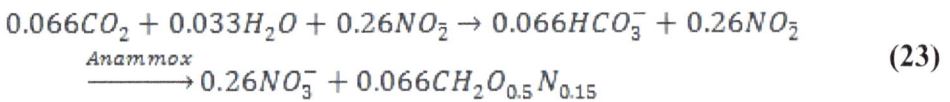

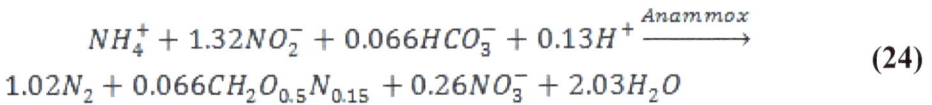

$$NH_4^+ + NO_2^- \xrightarrow{Anammox} N_2 + 2H_2O \tag{22}$$

$$0.066CO_2 + 0.033H_2O + 0.26NO_2^- \rightarrow 0.066HCO_3^- + 0.26NO_2^-$$
$$\xrightarrow{Anammox} 0.26NO_3^- + 0.066CH_2O_{0.5}N_{0.15} \tag{23}$$

$$NH_4^+ + 1.32NO_2^- + 0.066HCO_3^- + 0.13H^+ \xrightarrow{Anammox}$$
$$1.02N_2 + 0.066CH_2O_{0.5}N_{0.15} + 0.26NO_3^- + 2.03H_2O \tag{24}$$

The main product of Anammox reaction is nitrogen. Considering the main reaction equation, the conversion ratio of ammonium to nitrite is 1: 32, and the conversion ratio of nitrite to nitrate is 1: 0.26. Anammox reactive bacteria are autotrophs. They use bicarbonate as the main carbon resource, and nitrite as the main electron acceptor [43, 44].

The Effective Factors on Anammox Process

Free Ammonia and Ammonium

Some authors indicated that free ammonia (FA), even in low concentrations, restricts Anammox process more than ammonium ion [45]. However, recent studies show that lower than 13-38 mg/L FA concentration has no effect on Anammox process [46]. Aktan *et al.* observed the effects of FA on Anammox bacteria. Their results reveal that activity of bacteria decreases up to 90% with an increase in FA concentration up to 190 mg/L [47].

Nitrite

High concentration of nitrite is restrictive for Anammox bacteria. Lotti *et al.* stated that in Anammox process, nitrite inhibition potential is higher than nitrous acid [48, 49].

Organic Materials

An organic carbon resource will create some competitions between heterotrophic denitrifying bacteria and Anammox bacteria [50]. This is caused by high growth rate of heterotrophic bacteria in the presence of organic carbon. In C: N of higher than 2, the heterotrophic bacteria can remove Anammox bacteria. Also, some kinds of organic materials, such as methanol, are toxic for Anammox bacteria and restrict the process [51].

Temperature and pH

Many studies have introduced the optimal 30°C to 40°C temperature range for Anammox process [52, 53]. Dosta *et al.* observed the highest Anammox activity in a temperature range of 35°C to 40°C [54]. Nitrite accumulation is evident in low temperatures. The optimal pH value for this process is nearly 8. Furthermore, both temperature and pH affect FA and FNA processes as the influential parameters of Anammox process [55].

Salinity

Salinity affects Anammox process. However, Anammox bacteria can grow in both fresh and sea water. Therefore, this process can hopefully be used in salty wastewater treatment, but salt has a threshold inhibitory for bacteria. For example, Jin *et al.*observed that 30 g/L NaCl can decrease Anammox process up to 67.5% [56]. Some authors indicated that 6 g/L NaCl have no effects on Annamox activity, while 7.5 g/L KCl and 7.1 g/L Na_2SO_4 restrict its activity. Also, 3-15 g/L NaCl salinity increases the formation of Anammox granular sludge [57].

Oxygen

Anammox bacteria are anaerobic. So oxygen restricts their activity. However, some researchers announced that low oxygen concentrations do not restrict Anammox activity [55].

Partial Nitrication

Partial nitrification process is caused by nitrite accumulation as an intermediate compound in both nitrification and denitrification processes. In this method, ammonia is partially oxidized to nitrite, then nitrite is denitrified to N_2 [58 - 61]. The reaction of this process is as follows:

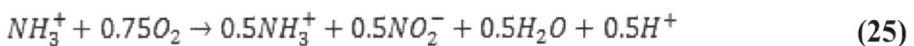

$$NH_3^+ + 0.75O_2 \rightarrow 0.5NH_3^+ + 0.5NO_2^- + 0.5H_2O + 0.5H^+ \qquad (25)$$

The tendency of nitrite oxidants to oxygen is higher than ammonia oxidants.

Therefore, with restricting nitrite oxidation without stopping ammonia oxidation, nitrite accumulation will be achieved. Some factors, such as dissolved oxygen, pH, and temperature, affect this process. So, with controlling these parameters, nitrite accumulation can be achieved. The results of some of the studies indicated that nitrite oxidation can be restricted by high ammonia concentrations. In high concentrations of FA, nitrite oxidation will be initially restricted. However, the bacteria will adopt to this amount of concentration. So, FA concentration should be increased gradually. Based on the above mentioned issues, it can be inferred that partial nitrification is effective for nitrogen-rich wastewaters [62 - 65].

Effective Parameters for Partial Nitrification

pH

pH is an important parameter for NOB activity inhibition and nitrite accumulation. In an alkali pH, nitrite accumulation is observable. Available reports indicated that the optimal pH for nitrite accumulation is 8.5. However, by controlling pH and keeping it in 8.5, nitritation process will be complete [66, 67].

DO

In terms of stoichiometry in a denitrification reaction, 3.43 mg of oxygen for 1 mg NH_3-N, and 1.14 mg for 1mg NO_2-N are used [68]. Low concentrations of dissolved oxygen are highly restrictive for nitrite oxidation [69]. Furthermore, minimum 0.3 mg/L DO concentration is required for nitrification [70]. Cecan *et al.*, indicated that, in a nitrification process, the ratio of oxygen to ammonia is more important than ammonia concentration in nitrite accumulation [71].

Temperature

Researchers proved the positive effects of temperature. Nitrite concentration and FA are higher in wastewater treatment facility in summers [72]. Gradual increase of temperature affects both AOB and NOB by forming NH_3 [73]. Furthermore, temperature increase up to 10°C will increase the rate of ammonia oxidation and nitrite up to 2.6 and 1.8, respectively. Specific growth rate of nitrobacter in the temperatures higher than 25°C is nearly equal to nitrosmonas causing nitrite accumulation [74].

Light

Light restricts both AOB and NOB. The rate of nitrification process is higher in dark places. The activity of nitrosmonas stops in 200 W light [75]. Furthermore, Oslon and Venzella observed that NOB sensitivity to sun light is higher than

AOB. Some authors reported that nitrification and nitration processes stop in summer due to high levels of light [76].

Simultaneous Nitrification Denitrification (SND)

Simultaneous nitrification denitrification (SND) can be carried out in special conditions. This process is considered to be an important process due to its effects on decreasing C: N ratio and anoxic volume [77]. Simultaneous nitrification denitrification in a reactor is important for decreasing the duration of nitrogen removal from wastewaters. A lot of studies are carried out on such kind of bacteria which can oxidize ammonia and decrease the amount of nitrate. Nitrification is an aerobic reaction and requires oxygen [78]:

$$3NH_4^+ + 6O_2 \rightarrow 3NO_3^- + 6H^+ + 3H_2O \tag{26}$$

Denitrification process is carried out in anoxic conditions:

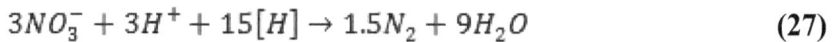

$$3NO_3^- + 3H^+ + 15[H] \rightarrow 1.5N_2 + 9H_2O \tag{27}$$

Some factors, such as oxygen concentration, carbon resource and floc size affect SND. Two mechanisms are reported for SND:

(1) the reason for concentration gradient created in floc is diffusionallimitation, and (2) The presence of aerobic denitrification along with heterotrophic nitrifying bacteria. The concentration gradient of DO in biofilm or flocs is due to diffusionallimitation [79]. The physical explanation for SND is that concentration gradient of DO is due to size and form of floc. The more access to the center of floc, the more the oxygen concentration will be. Therefore, in the center of floc, anoxic microorganism is evident. They produce nitrogen along with heterotrophic denitrifies. At the first process, denitrification starts with denitrifying microorganism in the first phase, then it moves toward a more aerobic condition. With increasing oxygen level, denitrification process gradually slows down and nitrification takes place at the same time. In the second mechanism, autotrophic microorganisms exist; nitrification and denitrification happen simultaneously [80]. Due to different conditions, reactions need to be controlled and maintained. However, some special kinds of bacteria can grow in this conditions [81]. The suitable conditions for nitrifying and denitrifying bacteria are similar to suitable conditions for SND process. Nitrosomonas and nitrobacter obtain the required energy from ammonium oxidation and nitrite oxidation, respectively. The oxidized nitrogen component can be used as an electron acceptor in denitrifying bacteria for nitrate removal [82]. *Nitosomonas europea* bacteria reduce nitrate and ammonium oxidation simultaneously and are very functional in SND processes

[83].

Multiple factors affect SND process, including: COD, DO, TN, HTR, SRT, carbon resource, and floc size. SND can be carried out in a DO carbon resource and a suitable floc size. DO gradient in floc is the most important factor. In SND processes in the external layer of floc with high oxygen concentration, nitrification is done by nitrifying bacteria. Denitrification can be carried out in the middle of the floc with totally anoxic condition. Based on the above mentioned theory, denitrification process is complete in the internal layer of floc due to lack of carbon resource. Therefore, the authors believe that by aeration, some carbon will be consumed from the outer environment, so denitrification process enhances in the internal layer of floc [84]. Some advantages and disadvantages of SND method are presented in Table **2**.

Table 2. advantages and disadvantages of SND

Advantages	Disadvantages
Removal output of 80 to 96 percent	Need for kinetic understanding
No need for a separate reactor for each process	Challenging design and control
Less nitrogen oxide generation	High SRT
Simple process design	Need for careful control of DO
Neutral pH	Hard control

Combined Partial Nitrification and Anammox Process

In this process, some parts of NH_4^+ oxidize into nitrite by aerobic AOB, and then, nitrite converts into N_2 gas. Other parts of NH_4^+ (by anaerobic bacteria) and NO_2^- oxidize into N_2; here NO_2^- acts as an electron acceptor. This combined process have some advantages such as 40% oxygen reduction in comparison with normal nitrogen removal systems, no need for organic material in denitrification process, and less sludge production [42]. This also has some disadvantages, such as:

1. Activity of anaerobic Anammox bacteria under the influence of remained DO in partial nitrification output which can be decreased.
2. Removal of optimal ratio of ammonium to nitrate (1: 1.3) is hard in this process.
3. The growth rate of AOB aerobic bacteria is higher than AOB anaerobic bacteria. Therefore, aerobic bacteria may become the dominant species in the output of partial nitrification.

CONCLUSION

Biological processes are efficient in removal of nitrogen compounds due to their effective function, high output, and less disadvantages in comparison with other methods. Furthermore, these processes have low costs and have no pre or post treatment steps in comparison with physico-chemical processes. In biological denitrification, nitrate converts into harmless nitrogen gas. Furthermore, in comparison with other methods, no more concentrated nitrate wastewater is generated. In the nitrification process, ammonium converts into nitrite and then to nitrate in a biological process. This process is taken place along with aerobic bacteria under the influence of the amount of dissolved oxygen. Also, other parameters such as temperature, ammonium level, pH, organic carbon and light affect the process. Simultaneous nitrification denitrification (SND) is a process in which nitrification and denitrification reactions take place simultaneously in a reactor. The main advantages of this process are its high output, low reactor volume, and low ratio of C: N. Sensitivity to oxygen and difficult control of the process are considered as its disadvantages. In partial nitrification process, ammonium oxidizes partially, and then, nitrite denitrifies into N_2. In this process, nitrite accumulation is used. The main advantage of this process is removal of extra oxidation conversion into nitrate. Its disadvantage is dependency on special and suitable condition for nitrite accumulation. Anammox process is one of the autotrophic and affordable processes in which Anammox bacteria, along with nitrite as an electron acceptor, converts ammonium into nitrogen. Some advantages of this process are: no need for aeration and carbon resource and low biomass yield. On the other hand, its disadvantages are nitrite supply and nitrite accumulation in the treated wastewater. The problem of nitrite supply is solved by combining partial nitrification with Anammox.

CONSENT FOR PUBLICATION

Not applicable.

REFERENCES

[1]   Halling-Sørensen B, Jorgensen SE. The removal of nitrogen compounds from wastewater. Elsevier 1993.

[2]   Hommes NG, Sayavedra-Soto LA, Arp DJ. Chemolithoorganotrophic growth of *Nitrosomonas Europaea* on fructose. J Bacteriol 2003; 185(23): 6809-14.
       [http://dx.doi.org/10.1128/JB.185.23.6809-6814.2003] [PMID: 14617645]

[3]   Grady CL Jr, Daigger GT, Love NG, Filipe CD. Biological wastewater treatment. CRC Press 2011.

[4]   Painter HA. Microbial transformations of inorganic nitrogen. In: Jenkins SH, Ed. Proceedings of the Conference on Nitrogen as a Water Pollutant: Pergamon 2013; p. 3-29.
       [http://dx.doi.org/10.1016/B978-1-4832-1344-6.50003-4]

[5]   Wolfe RL, Lieu NI. Nitrifying bacteria in drinking water. In: Bitton G, Ed. In Encyclopedia of

Environmental Microbiology 2003.
[http://dx.doi.org/10.1002/0471263397.env228]

[6]   Grunditz C, Dalhammar G. Development of nitrification inhibition assays using pure cultures of Nitrosomonas and Nitrobacter. Water Res 2001; 35(2): 433-40.
[http://dx.doi.org/10.1016/S0043-1354(00)00312-2] [PMID: 11228996]

[7]   Alleman JE, Keramida V, Pantea-Kiser L. Light induced Nitrosomonas inhibition. Water Res 1987; 21(4): 499-501.
[http://dx.doi.org/10.1016/0043-1354(87)90199-0]

[8]   Odell LH, Kirmeyer GJ, Wilczak A, *et al.* Controlling nitrification in chloraminated systems. J Am Water Works Assoc 1996; 88(7): 86-98.
[http://dx.doi.org/10.1002/j.1551-8833.1996.tb06587.x]

[9]   Wilczak A, Jacangelo JG, Marcinko JP, Odell LH, Kirmeyer GJ. Occurrence of nitrification in chloraminated distribution systems. J Am Water Works Assoc 1996; 88(7): 74-85.
[http://dx.doi.org/10.1002/j.1551-8833.1996.tb06586.x]

[10]   Hockenbury MR, Grady C, Daigger GT. Factors affecting nitrification. J Environ Eng 1977; 103(1): 9-19.
[http://dx.doi.org/10.1080/00103620802004235]

[11]   Sharma B, Ahlert R. Nitrification and nitrogen removal. Water Res 1977; 11(10): 897-925.
[http://dx.doi.org/10.1016/0043-1354(77)90078-1]

[12]   Sattley Sattley WM, Madigan MT. Isolation, Characterization, and Ecology of Cold-Active, Chemolithotrophic, Sulfur-Oxidizing Bacteria from Perennially Ice-Covered Lake Fryxell, Antarctica. Appl Environ Microbiol 2006; 72(8): 5562-68.
[http://dx.doi.org/10.1128/AEM.00702-06]

[13]   Morton SC, Zhang Y, Edwards MA. Implications of nutrient release from iron metal for microbial regrowth in water distribution systems. Water Res 2005; 39(13): 2883-92.
[http://dx.doi.org/10.1016/j.watres.2005.05.024] [PMID: 16029882]

[14]   Fransolet G, Depelchin A, Villers G, Goossens R, Masschelein WJ. The role of bicarbonate in bacterial growth in oligotrophic waters. J Am Water Works Assoc 1988; 80(11): 57-61.
[http://dx.doi.org/10.1002/j.1551-8833.1988.tb03135.x]

[15]   Mariotti A. Denitrification in groundwaters, principles and methods for its identification-A review. J Hydrol 1986; 88(1-2): 1-23.
[http://dx.doi.org/10.1080/10934529609376448]

[16]   Lee K-C, Rittmann BE. Applying a novel autohydrogenotrophic hollow-fiber membrane biofilm reactor for denitrification of drinking water. Water Res 2002; 36(8): 2040-52.
[http://dx.doi.org/10.1016/S0043-1354(01)00425-0] [PMID: 12092579]

[17]   Metcalf L, Eddy HP. Wastewater engineering collection treatment disposal: McGraw-Hill Interamericana 1972.

[18]   Abu-Ghararah ZH. Biological denitrification of high nitrate water: Influence of type of carbon source and nitrate loading. J Environ Sci Health A 1996; 31(7): 1651-68.
[http://dx.doi.org/10.1080/10934529609376448]

[19]   Rogalla F, de Larminat G, Coutelle J, Godart H. Experience with Nitrate-Removal Methods from Drinking Water.Nitrate Contamination NATO ASI Series (Series G: Ecological Sciences). Berlin, Heidelberg: Springer 1991; Vol. 30.
[http://dx.doi.org/10.1007/978-3-642-76040-2_27]

[20]   Matějů V, Čižinská S, Krejčí J, Janoch T. Biological water denitrification—a review. Enzyme Microb Technol 1992; 14(3): 170-83.
[http://dx.doi.org/10.1016/0141-0229(92)90062-S]

[21] Rott U, Lamberth B. Subterranean denitrification for the treatment of drinking water. Water Supply 1992; 10(3): 111-20.

[22] Rivett MO, Buss SR, Morgan P, Smith JW, Bemment CD. Nitrate attenuation in groundwater: A review of biogeochemical controlling processes. Water Res 2008; 42(16): 4215-32.
[http://dx.doi.org/10.1016/j.watres.2008.07.020] [PMID: 18721996]

[23] Chang CC, Tseng SK, Huang HK. Hydrogenotrophic denitrification with immobilized *Alcaligenes eutrophus* for drinking water treatment. Bioresour Technol 1999; 69(1): 53-8.
[http://dx.doi.org/10.1016/S0960-8524(98)00168-0]

[24] Terada A, Kaku S, Matsumoto S, Tsuneda S. Rapid autohydrogenotrophic denitrification by a membrane biofilm reactor equipped with a fibrous support around a gas-permeable membrane. Biochem Eng J 2006; 31(1): 84-91.
[http://dx.doi.org/10.1016/j.bej.2006.06.004]

[25] Balows A, Truper H, Dworkin M, Harder W, Schleifer K, Eds. A Handbook on the Biology of Bacteria: Ecophysiology, Isolation, Identification, Applications. New York: Publisher-Springer-Verlag 1992; LXXXVI: p. 1155.

[26] Canter LW. Nitrates in Groundwater. Florida, USA: CRC, Boca Raton 1997; p. 288.

[27] Park HI. kun Kim D, Choi Y-J, Pak D. Nitrate reduction using an electrode as direct electron donor in a biofilm-electrode reactor. Process Biochem 2005; 40(10): 3383-8.
[http://dx.doi.org/10.1016/j.procbio.2005.03.017]

[28] Zhou M, Fu W, Gu H, Lei L. Nitrate removal from groundwater by a novel three-dimensional electrode biofilm reactor. Electrochim Acta 2007; 52(19): 6052-9.
[http://dx.doi.org/10.1016/j.electacta.2007.03.064]

[29] Lee K-C, Rittmann BE. Effects of pH and precipitation on autohydrogenotrophic denitrification using the hollow-fiber membrane-biofilm reactor. Water Res 2003; 37(7): 1551-6.
[http://dx.doi.org/10.1016/S0043-1354(02)00519-5] [PMID: 12600383]

[30] Lu CX, Gu P. Hydrogenotrophic denitrification for the removal of nitrate in drinking water. Huan Jing Ke Xue 2008; 29(3): 671-6.
[PMID: 18649526]

[31] Ghafari S, Hasan M, Aroua MK. Improvement of autohydrogenotrophic nitrite reduction rate through optimization of pH and sodium bicarbonate dose in batch experiments. J Biosci Bioeng 2009; 107(3): 275-80.
[http://dx.doi.org/10.1016/j.jbiosc.2008.11.008] [PMID: 19269592]

[32] Haugen KS, Semmens MJ, Novak PJ. A novel *in situ* technology for the treatment of nitrate contaminated groundwater. Water Res 2002; 36(14): 3497-506.
[http://dx.doi.org/10.1016/S0043-1354(02)00043-X] [PMID: 12230195]

[33] Brady NC, Weil RR. The Nature and Properties of Soils.15th Edition, Pearson Education Limited, Inc, Upper Saddle River 2017; xvii: p. 1086.

[34] Rezania B, Cicek N, Oleszkiewicz JA. Kinetics of hydrogen-dependent denitrification under varying pH and temperature conditions. Biotechnol Bioeng 2005; 92(7): 900-6.
[http://dx.doi.org/10.1002/bit.20664] [PMID: 16116656]

[35] Dries D, Liessens J, Verstraete W, *et al.* Nitrate removal from drinking water by means of hydrogenotrophic denitrifiers in a polyurethane carrier reactor. Water Supply 1988; 6: 181-92.

[36] Sakakibara Y, Nakayama T. A novel multi-electrode system for electrolytic and biological water treatments: electric charge transfer and application to denitrification. Water Res 2001; 35(3): 768-78.
[http://dx.doi.org/10.1016/S0043-1354(00)00327-4] [PMID: 11228976]

[37] Karanasios K, Michailides M, Vasiliadou I, Pavlou S, Vayenas D. Potable water hydrogenotrophic denitrification in packed-bed bioreactors coupled with a solar-electrolysis hydrogen production

system. Desalination and Water Treatment 2011; 33(1-3): 86-96.
[http://dx.doi.org/10.5004/dwt.2011.2614]

[38]   Kim Y-S, Nakano K, Lee T-J, Kanchanatawee S, Matsumura M. On-site nitrate removal of groundwater by an immobilized psychrophilic denitrifier using soluble starch as a carbon source. J Biosci Bioeng 2002; 93(3): 303-8.
[http://dx.doi.org/10.1016/S1389-1723(02)80032-9] [PMID: 16233204]

[39]   Nair RR, Dhamole PB, Lele SS, D'Souza SF. Biological denitrification of high strength nitrate waste using preadapted denitrifying sludge. Chemosphere 2007; 67(8): 1612-7.
[http://dx.doi.org/10.1016/j.chemosphere.2006.11.043] [PMID: 17234243]

[40]   Mulder A, Van de Graaf AA, Robertson L, Kuenen J. Anaerobic ammonium oxidation discovered in a denitrifying fluidized bed reactor. FEMS Microbiol Ecol 1995; 16(3): 177-83.
[http://dx.doi.org/10.1111/j.1574-6941.1995.tb00281.x]

[41]   Mulder A. assignee. Anoxic ammonia oxidation. United States patent US 5,078,884, 1992 Jan 7;

[42]   Jetten MS, Horn SJ, van Loosdrecht MC. Towards a more sustainable municipal wastewater treatment system. Water Sci Technol 1997; 35(9): 171-80.
[http://dx.doi.org/10.2166/wst.1997.0341]

[43]   Third KA, Sliekers AO, Kuenen JG, Jetten MS. The CANON system (Completely Autotrophic Nitrogen-removal Over Nitrite) under ammonium limitation: interaction and competition between three groups of bacteria. Syst Appl Microbiol 2001; 24(4): 588-96.
[http://dx.doi.org/10.1078/0723-2020-00077] [PMID: 11876366]

[44]   van der Star WR, Miclea AI, van Dongen UG, Muyzer G, Picioreanu C, van Loosdrecht MC. The membrane bioreactor: a novel tool to grow anammox bacteria as free cells. Biotechnol Bioeng 2008; 101(2): 286-94.
[http://dx.doi.org/10.1002/bit.21891] [PMID: 18421799]

[45]   Jung JY, Kang SH, Chung YC, Ahn DH. Factors affecting the activity of anammox bacteria during start up in the continuous culture reactor. Water Sci Technol 2007; 55(1-2): 459-68.
[http://dx.doi.org/10.2166/wst.2007.023] [PMID: 17305171]

[46]   Waki M, Tokutomi T, Yokoyama H, Tanaka Y. Nitrogen removal from animal waste treatment water by anammox enrichment. Bioresour Technol 2007; 98(14): 2775-80.
[http://dx.doi.org/10.1016/j.biortech.2006.09.031] [PMID: 17092710]

[47]   Aktan CK, Yapsakli K, Mertoglu B. Inhibitory effects of free ammonia on Anammox bacteria. Biodegradation 2012; 23(5): 751-62.
[http://dx.doi.org/10.1007/s10532-012-9550-0] [PMID: 22460564]

[48]   Lotti T, van der Star WR, Kleerebezem R, Lubello C, van Loosdrecht MC. The effect of nitrite inhibition on the anammox process. Water Res 2012; 46(8): 2559-69.
[http://dx.doi.org/10.1016/j.watres.2012.02.011] [PMID: 22424965]

[49]   Jaroszynski LW, Cicek N, Sparling R, Oleszkiewicz JA. Importance of the operating pH in maintaining the stability of anoxic ammonium oxidation (anammox) activity in moving bed biofilm reactors. Bioresour Technol 2011; 102(14): 7051-6.
[http://dx.doi.org/10.1016/j.biortech.2011.04.069] [PMID: 21565492]

[50]   Molinuevo B, García MC, Karakashev D, Angelidaki I. Anammox for ammonia removal from pig manure effluents: effect of organic matter content on process performance. Bioresour Technol 2009; 100(7): 2171-5.
[http://dx.doi.org/10.1016/j.biortech.2008.10.038] [PMID: 19097886]

[51]   Kumar M, Lin J-G. Co-existence of anammox and denitrification for simultaneous nitrogen and carbon removal--Strategies and issues. J Hazard Mater 2010; 178(1-3): 1-9.
[http://dx.doi.org/10.1016/j.jhazmat.2010.01.077] [PMID: 20138428]

[52]   Egli K, Fanger U, Alvarez PJ, Siegrist H, van der Meer JR, Zehnder AJ. Enrichment and

characterization of an anammox bacterium from a rotating biological contactor treating ammonium-rich leachate. Arch Microbiol 2001; 175(3): 198-207.
[http://dx.doi.org/10.1007/s002030100255] [PMID: 11357512]

[53]   Strous M, Heijnen J, Kuenen JG, Jetten M. The sequencing batch reactor as a powerful tool for the study of slowly growing anaerobic ammonium-oxidizing microorganisms. Appl Microbiol Biotechnol 1998; 50(5): 589-96.
[http://dx.doi.org/10.1007/s002530051340]

[54]   Dosta J, Fernández I, Vázquez-Padín JR, *et al.* Short- and long-term effects of temperature on the Anammox process. J Hazard Mater 2008; 154(1-3): 688-93.
[http://dx.doi.org/10.1016/j.jhazmat.2007.10.082] [PMID: 18063297]

[55]   Strous M, Van Gerven E, Zheng P, Kuenen JG, Jetten MS. Ammonium removal from concentrated waste streams with the anaerobic ammonium oxidation (anammox) process in different reactor configurations. Water Res 1997; 31(8): 1955-62.
[http://dx.doi.org/10.1016/S0043-1354(97)00055-9]

[56]   Jin R-C, Ma C, Mahmood Q, Yang G-F, Zheng P. Anammox in a UASB reactor treating saline wastewater. Process Saf Environ Prot 2011; 89(5): 342-8.
[http://dx.doi.org/10.1016/j.psep.2011.05.001]

[57]   Dapena-Mora A, Vázquez-Padín J, Campos J, *et al.* Monitoring the stability of an Anammox reactor under high salinity conditions. Biochem Eng J 2010; 51(3): 167-71.
[http://dx.doi.org/10.1016/j.bej.2010.06.014]

[58]   Rittmann BE, McCarty PL. Environmental biotechnology: principles and applications: Tata McGraw-Hill Education 2012.
[http://dx.doi.org/10.1016/j.bej.2010.06.014]

[59]   Feng Y-J, Tseng S-K, Hsia T-H, Ho C-M, Chou W-P. Partial nitrification of ammonium-rich wastewater as pretreatment for anaerobic ammonium oxidation (Anammox) using membrane aeration bioreactor. J Biosci Bioeng 2007; 104(3): 182-7.
[http://dx.doi.org/10.1263/jbb.104.182] [PMID: 17964481]

[60]   Hooper A. Biochemistry of the nitrifying lithoautotrophic bacteria. Autotrophic bacteria 1989; 239-65.

[61]   Chung J, Bae W. Nitrite reduction by a mixed culture under conditions relevant to shortcut biological nitrogen removal. Biodegradation 2002; 13(3): 163-70.
[http://dx.doi.org/10.1023/A:1020896412365] [PMID: 12498214]

[62]   Turk O, Mavinic DS. Benefits of using selective inhibition to remove nitrogen from highly nitrogenous wastes. Environ Technol Lett 1987; 8(1-12): 419-26.
[http://dx.doi.org/10.1080/09593338709384500]

[63]   Beccari M, Passino R, Ramadori R, Tandoi V. Kinetics of dissimilatory nitrate and nitrite reduction in suspended growth culture. Water Pollut Control Fed 1983; pp. 58-64.

[64]   Bae W, Baek S, Chung J, Lee Y. Nitrite accumulation in batch reactor under various operational conditions. Biodegradation 2002; 12: 359-66.
[http://dx.doi.org/10.1023/A:1014308229656] [PMID: 11995828]

[65]   Abeling U, Seyfried C. Anaerobic-aerobic treatment of high-strength ammonium wastewater-nitrogen removal *via* nitrite. Water Sci Technol 1992; 26(5-6): 1007-15.
[http://dx.doi.org/10.2166/wst.1992.0542]

[66]   Balmelle B, Nguyen K, Capdeville B, Cornier J, Deguin A. Study of factors controlling nitrite build-up in biological processes for water nitrification. Water Sci Technol 1992; 26(5-6): 1017-25.
[http://dx.doi.org/10.2166/wst.1992.0543]

[67]   Tokutomi T. Operation of a nitrite-type airlift reactor at low DO concentration. Water Sci Technol 2004; 49(5-6): 81-8.
[http://dx.doi.org/10.2166/wst.2004.0740] [PMID: 15137410]

[68]   Goreau TJ, Kaplan WA, Wofsy SC, McElroy MB, Valois FW, Watson SW. Production of NO_2^- and N_2O by nitrifying bacteria at reduced concentrations of oxygen. Appl Environ Microbiol 1980; 40(3): 526-32.
        [PMID: 16345632]

[69]   Hanaki K, Wantawin C, Ohgaki S. Nitrification at low levels of dissolved oxygen with and without organic loading in a suspended-growth reactor. Water Res 1990; 24(3): 297-302.
        [http://dx.doi.org/10.1016/0043-1354(90)90004-P]

[70]   Stenstrom MK, Poduska RA. The effect of dissolved oxygen concentration on nitrification. Water Res 1980; 14(6): 643-9.
        [http://dx.doi.org/10.1016/0043-1354(80)90122-0]

[71]   Çeçen F, Gönenç IE. Criteria for nitrification and denitrification of high-strength wastes in two upflow submerged filters. Water Environ Res 1995; 67(2): 132-42.
        [http://dx.doi.org/10.2175/106143095X131277]

[72]   Gelda RK, Effler SW, Brooks CM. Nitrite and the two stages of nitrification in nitrogen polluted Onondaga Lake, New York. J Environ Qual 1999; 28(5): 1505-17.
        [http://dx.doi.org/10.2134/jeq1999.00472425002800050016x]

[73]   Fdz-Polanco F, Villaverde S, Garcia P. Temperature effect on nitrifying bacteria activity in biofilters: activation and free ammonia inhibition. Water Sci Technol 1994; 30(11): 121-30.
        [http://dx.doi.org/10.2166/wst.1994.0552]

[74]   Knowles G, Downing AL, Barrett MJ. Determination of kinetic constants for nitrifying bacteria in mixed culture, with the aid of an electronic computer. J Gen Microbiol 1965; 38(2): 263-78.
        [http://dx.doi.org/10.1099/00221287-38-2-263] [PMID: 14287204]

[75]   Hooper AB, Terry KR. Specific inhibitors of ammonia oxidation in Nitrosomonas. J Bacteriol 1973; 115(2): 480-5.
        [PMID: 4725614]

[76]   Olson RJ. Differential photoinhibition of marine nitrifying bacteria: a possible mechanism for the formation of the primary nitrite maximum. J Mar Res 1981; 39: 227-38.

[77]   Zhu G, Peng Y, Li B, Guo J, Yang Q, Wang S. Biological Removal of Nitrogen from Wastewater.Reviews of Environmental Contamination and Toxicology Reviews of Environmental Contamination and Toxicology. New York, NY: Springer 2008; Vol. 192.
        [http://dx.doi.org/10.1007/978-0-387-71724-1_5]

[78]   Casey E, Glennon B, Hamer G. Review of membrane aerated biofilm reactors. Resour Conserv Recycling 1999; 27(1-2): 203-15.
        [http://dx.doi.org/10.1016/S0921-3449(99)00007-5]

[79]   Zhu GB, Peng YZ, Wu SY, Wang SY, Xu SW. Simultaneous nitrification and denitrification in step feeding biological nitrogen removal process. J Environ Sci (China) 2007; 19(9): 1043-8.
        [http://dx.doi.org/10.1016/S1001-0742(07)60170-3] [PMID: 17966507]

[80]   Gogina E, Gulshin I. Simultaneous nitrification and denitrification with low dissolved oxygen level and C/N ratio. Procedia Eng 2016; 153: 189-94.
        [http://dx.doi.org/10.1016/j.proeng.2016.08.101]

[81]   Schmidt I, Sliekers O, Schmid M, *et al.* New concepts of microbial treatment processes for the nitrogen removal in wastewater. FEMS Microbiol Rev 2003; 27(4): 481-92.
        [http://dx.doi.org/10.1016/S0168-6445(03)00039-1] [PMID: 14550941]

[82]   Sliekers AO, Derwort N, Gomez JL, Strous M, Kuenen JG, Jetten MS. Completely autotrophic nitrogen removal over nitrite in one single reactor. Water Res 2002; 36(10): 2475-82.
        [http://dx.doi.org/10.1016/S0043-1354(01)00476-6] [PMID: 12153013]

[83]   Shrestha NK, Hadano S, Kamachi T, Okura I. Dinitrogen production from ammonia by *Nitrosomonas*

europaea. Appl Catal A 2002; 237(1-2): 33-9.
[http://dx.doi.org/10.1016/S0926-860X(02)00279-X]

[84]   Puznava N, Payraudeau M, Thornberg D. Simultaneous nitrification and denitrification in biofilters with real time aeration control. Water Sci Technol 2001; 43(1): 269-76.
[http://dx.doi.org/10.2166/wst.2001.0057] [PMID: 11379100]

CHAPTER 3

Conventional and Novel Biological Nitrogen Removal Processes for Nitrogen Removal from Food Industries Wastewater

Ghazaleh Mirbolouki Tochaei[1], Mehrdad Farrokhi[2], Mehrdad Moslemzadeh[1], Saeid Ildari[1] and Mostafa Mahdavianpour[3,*]

[1] *Department of Environmental Health Engineering, School of Health, Guilan University of Medical Sciences, Rasht, Iran*

[2] *Research Center for Health in Disasters and Emergencies, University of Social Welfare and Rehabilitation Sciences, Tehran, Iran*

[3] *Department of Environmental Health Engineering, Abadan School of Medical Sciences, Abadan, Iran*

Abstract: Normally, the food industry produces wastewater containing high organic matter and nitrogen compounds. Along with organic matter removal from these streams, also, nitrogen compounds should be removed. Nitrogen can be removed biologically from wastewater using conventional and novel processes. Although, conventional nitrification/denitrification is an established system for nitrogen removal but the costs of treatment by this system are high because of: high oxygen requirement for nitrification, the addition of external carbon source for denitrification (in the case of wastewaters with low C/N ratio), slow growth rate of microorganisms responsible for both processes, and needing two successive reactor or difficult control of one-reactor system. These disadvantages led to conduction of substantial studies to find better alternatives. Therefore, different novel processes were studied and used for nitrogen removal. Many autotrophic processes including SHARON, ANAMMOX, SHARON-ANAMMOX, CANON, and OLAND were discovered to be more economical than conventional nitrification/denitrification system. Also, more studies on combination of novel processes with a part of conventional nitrification/denitrification process led to the development of other novel processes such as heterotrophic nitrification-aerobic denitrification, NO_x, and DEAMOX processes. Details of both conventional and novel processes were discussed in this chapter. Finally, the possibility of using these processes to remove nitrogen from food industry wastewater is discussed. Because of co-existence of both carbon and nitrogen compounds in food industry wastewater, the use of novel autotrophic processes for mainstream wastewater treatment is impossible. So, carbonaceous contaminants should be removed using anaerobic digestion or high load conventional activated sludge, and then effluent of these processes with high nitrogen contents (low C/N ratio) can be treated using novel autotrophic processes. It

---

*****Corresponding author Mostafa Mahdavianpour:** Department of Environmental Health Engineering, Abadan School of Medical Sciences, Abadan, Iran; E-mail: mmp368@gmail.com

Edris Hoseinzadeh (Ed.)

should be noted that there are two alternatives including heterotrophic nitrification-aerobic denitrification process and organic-laden DEAMOX process that can be used for mainstream food industry wastewater treatment but the exact characteristics of these processes are unclear and should be accurately studied in laboratory, pilot and full-scale before use.

Keywords: Aerobic Deammonification, ANAMMOX Process, CANON Process, Conventional Nitrification/denitrification, DEAMOX Process, Food Industry, Heterotrophic Nitrification-aerobic Denitrification, Nitrogen Removal, NO_X Process, OLAND Process, SHARON Process.

INTRODUCTION

Wastewater streams resulted from food industry are generally characterized as rich wastewaters in the case of organic and nutrient contents and tend to destroy water environment if discharged without proper treatment [1 - 6]. Characterization of wastewaters from different food processing industries is shown in Table **1**. As can be seen, both organic and nitrogen contents of food industries are very high and make wastewaters produced in these industries as high-strength wastewaters. Reported by [1 - 6].

Release of nitrogen into water resources has severe environmental problems. It can be toxic to aquatic organisms and cause oxygen consumption and algal bloom (eutrophication) in receiving waters. Also, some of the nitrogen-containing compounds can consume chlorine during disinfection. Eutrophication can destroy water resources due to receiving high loads of nutrients. Eutrophication lowers water quality, changes the ecological body and function of water resources and poses many potential threats to human and animal life. The necessity of environmental and human health protection has led to the establishment of continuously increasing wastewater treatment plant (WWTP) discharge standards for nutrients and has motivated efforts to obtain better effluent quality, especially related to nitrogen. Nitrogenous compounds in wastewaters can be divided into two classes of nonbiodegradable and biodegradable. The nonbiodegradable nitrogenous class is associated with the nonbiodegradable particulate organic matters. The biodegradable nitrogenous class is subdivided into three parts of ammonium (NH_4^+), soluble organic nitrogen and particulate organic nitrogen. Particulate organic nitrogen is finally hydrolyzed to soluble organic nitrogen. Different physicochemical and biological processes are used to remove nitrogen compounds from wastewater. Biological processes are preferred to physico-chemical processes because of the efficient transformation of fixed nitrogenous compounds to harmless dinitrogen gas (N_2) in a cost-effective way [7 - 11]. Biological nitrogen removal processes are subdivided into conventional nitrification-denitrification and novel processes. The mechanisms and performances of all the processes are described in detail below.

Table 1. Characterization of wastewater from different food processing industries.

Industry	COD (mg/L)	Total N (mg/L)	NH_4^+-N (mg/L)
Food processing	1500-2000	350-500	300-450
Slaughterhouse	1000-6000	100-700	65-500
Dairy	1000-61000	50-1462	8-510
Potato-processing	5250	225	55
Fish canning	1147-8313	21-471 (soluble)	3-250

CONVENTIONAL NITRIFICATION-DENITRIFICATION

Conventional biological nitrogen removal includes two steps of aerobic nitrification and anoxic denitrification. In nitrification, ammonia chemolithoautotrophically oxidizes to nitrate under strictly aerobic conditions and this step itself is done in two sequential oxidative steps of ammonium to nitrite (ammonium oxidation) and nitrite to nitrate (nitrite oxidation), respectively. Each step of nitrification is performed by a different bacterial group which uses ammonia or nitrite as an energy source and molecular oxygen as a final electron acceptor, while carbon dioxide is used as the carbon source. In the first step of nitrification, ammonium (NH_4^+) oxidizes to nitrite (NO_2^-) through hydroxylamine (NH_2OH) path by ammonia-oxidizing bacteria (AOB). Oxidation of ammonia to hydroxylamine is catalyzed by membrane-bound ammonia monooxygenase (AMO). Molecular oxygen and dinitrogen tetroxide (N_2O_4, the dimer of NO_2) are the most likely electron acceptors for AMO (Eqs.1 and 2). Hydroxylamine oxidoreductase (HAO) further oxidizes ammonia to nitrite (Eq. (3)) [7, 11].

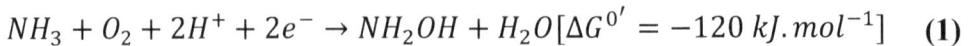

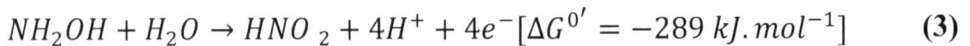

$$NH_3 + O_2 + 2H^+ + 2e^- \rightarrow NH_2OH + H_2O [\Delta G^{0'} = -120 \; kJ.mol^{-1}] \quad \textbf{(1)}$$

$$NH_3 + N_2O_4 + 2H^+ + 2e^- \rightarrow NH_2OH + 2NO + H_2O \; [\Delta G^{0'} = -140 \; kJ.mol^{-1}] \quad \textbf{(2)}$$

$$NH_2OH + H_2O \rightarrow HNO_2 + 4H^+ + 4e^- [\Delta G^{0'} = -289 \; kJ.mol^{-1}] \quad \textbf{(3)}$$

The four electrons released from reaction (3) enter the AMO reactions (Eqs.1 and 2), the CO_2 assimilation and the respiratory path. The released electrons are transferred to the final electron acceptors (oxygen in aerobic conditions). In the aerobic conditions, the oxygen oxidizes NO to NO_2. So, only small concentrations of NO are released to the emitted gas. Based on Eq. (2), dinitrogen tetroxide is the oxidizing agent even under aerobic conditions. In this reaction, hydroxylamine and NO are produced as by-products. Finally, hydroxylamine and NO are then oxidized to nitrite (Eq. (3)) and (N_2O_4) (Eq. (4)), respectively [9, 11, 12].

$$2NO + O_2 \rightarrow NO_2(N_2O_4) \tag{4}$$

Nitrosomonas is known as the most common AOB. Other AOB are *Nitrosococcus*, *Nitrosopira*, *Nitrosovibrio* and *Nitrosolobus*. These AOB are genetically different, but they are placed in the beta sub-division of the *Proteobacteria*. In the second step of nitrification, nitrite oxidizes to nitrate (NO_3^-) by nitrite-oxidizing bacteria (NOB). Membrane-bound nitrite oxidoreductase (NOR) catalyzes this reaction, in which nitrite oxidizes to nitrate with oxygen as the final electron acceptor (Eq. (5)) [7, 9, 11].

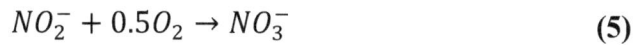

$$NO_2^- + 0.5O_2 \rightarrow NO_3^- \tag{5}$$

In this step, several genera such as *Nitrococcus*, *Nitrospira* and *Nitrocystis* are known to conduct this process. However, the most common NOB genus is *Nitrobacter*. *Nitrobacter* is also tightly linked within the α-*Proteobacteria*. Considering cell synthesis, the reactions occurred during nitrification are as follows [7, 9, 11, 12]:

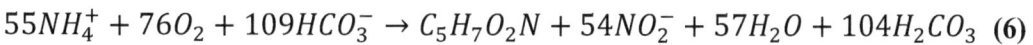

$$55NH_4^+ + 76O_2 + 109HCO_3^- \rightarrow C_5H_7O_2N + 54NO_2^- + 57H_2O + 104H_2CO_3 \tag{6}$$

$$400NO_2^- + NH_4^+ + 4H_2CO_3 + 195O_2 + HCO_3^- \rightarrow C_5H_7O_2N + +3H_2O + 400NO_3^- \tag{7}$$

The sum of the above reactions ((6) and (7)) can be represented as follows:

$$NH_4^+ + 1.37O_2 + 1.98HCO_3^- \rightarrow 0.02C_5H_7O_2N + 0.98NO_2^- + 1.04H_2O + 1.88\,H_2CO_3 \tag{8}$$

In these reactions, cell yields (Y) for *Nitrosomonas* and *Nitrobacter* are 0.15 mg cells/mg NH_4^+-N oxidized and 0.02 mg cells/mg NO_2^--N oxidized, respectively. In the equations, the ratio of consumed oxygen to oxidized substrates are 3.16 $mgO_2/mgNH_4^+$-N and 1.11 $mgO_2/mgNO_2^-$-N, respectively. Also, in these reactions, 7.07 mg alkalinity as $CaCO_3$ is required for the oxidation of each mg of NH_4^+-N. However, severe pH drop can occur when alkalinity in the wastewater is neutralized by the produced acid in the nitrification process. Following the pH drop (below 7.0) in the nitrification process, the reaction rates are rapidly decreased. Therefore, when the alkalinity of the wastewater is not enough to prevent pH drop resulted from the produced acid by nitrification, the external alkalinity source, such as lime, must be added. During denitrification, nitrate produced in nitrification is biologically reduced to gaseous N_2 under anoxic condition. Nitrite and nitrate are reduced to N_2 by heterotrophic microorganisms that use these substrates instead of oxygen as electron acceptors and organic matter (instead of CO_2) as carbon and energy source. In the denitrification, the denitrifying bacteria reduce nitrate to nitrite by catalyzing enzyme nitrate reductase (Nar). Then, nitrite is converted into nitrous oxide by nitrite reductase (Nir). Produced nitrous oxide is transformed into nitric oxide by nitric oxide reductase (Nor) and, finally, nitric oxide is reduced to nitrogen gas by nitrous oxide reductase (Nos). Denitrification is known as a type of respiration that occurs in the anoxic condition. Electrons originated from electron donors such as organic matter (OM), reduced sulfur compounds or molecular hydrogen in the absence of oxygen are transferred to the oxidized nitrogen compounds (NO_3^- and NO_2^-) to build up a proton motive force (PMF) necessary for ATP production. The involved enzymes are the Nar, the Nir, the Nor, and finally, the Nos. The main final product of denitrification is N_2, while NO and N_2O are produced as intermediates at low concentrations. However, these intermediates can be emitted as final products when, because of the inhibitory condition involved, the enzymes are not completely expressed, *e.g.* in the condition that the concentration of dissolved oxygen is too high. Denitrifiers are common among the Gram-negative α and β classes of the *Proteobacteria*. These classes include *Pseudomonas*, *Alcaligenes*, *Paracoccus* and *Thiobacillus*. Also, some Gram-positive bacteria (such as *Bacillus*) and a few halophilic *Archaea* (such as *Halobacterium*) can reduce oxidized nitrogen compounds to N_2. The denitrification can be accomplished by a variety of electron donors and carbon sources such as methanol, glucose, acetate, ethanol and a few others. Methanol (CH_3OH), with relatively low cost, is the most common external electron donor for denitrification. Using methanol as an electron donor, the overall dissimilation-synthesis reactions for denitrification are as follows:

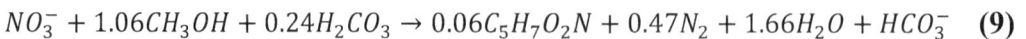

$$NO_3^- + 1.06CH_3OH + 0.24H_2CO_3 \rightarrow 0.06C_5H_7O_2N + 0.47N_2 + 1.66H_2O + HCO_3^- \quad (9)$$

In these reactions, theoretically, 2.41 mg of methanol are needed to reduce each milligram of NO_3-N. Without considering synthesis in the reaction, the required methanol is decreased to 1.9 milligram. Also, for the calculation of methanol needed for nitrite reduction and deoxygenation, Eqs. (10) and (11) can be used [7, 9 - 12].

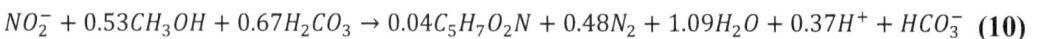

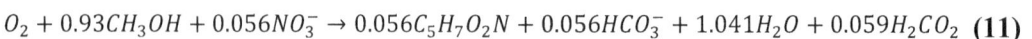

$$NO_2^- + 0.53CH_3OH + 0.67H_2CO_3 \rightarrow 0.04C_5H_7O_2N + 0.48N_2 + 1.09H_2O + 0.37H^+ + HCO_3^- \quad (10)$$

$$O_2 + 0.93CH_3OH + 0.056NO_3^- \rightarrow 0.056C_5H_7O_2N + 0.056HCO_3^- + 1.041H_2O + 0.059H_2CO_2 \quad (11)$$

NITROGEN REMOVAL FROM FOOD INDUSTRY USING CONVENTIONAL NITRIFICATION-DENITRIFICATION

Because of the co-existence of high organic and nitrogen contents in food industries, it seems that conventional nitrification-denitrification is an appropriate system to treat these wastewaters. As seen in Table **2**, different systems were applied to treat a variety of food industries wastewaters with a broad range of organic and nitrogen contents and all the applied processes had significant performances for both organic and nitrogen contaminants.

Table 2. Removal of organics and nitrogen compounds in conventional nitrification-denitrification using different bioreactors.

System	Wastewater characteristics (mg/L)	Organics removal (COD or TOC) (%)	NH_4^+ removal (%)	TN removal (%
Anoxic reactor/aerobic MBR	COD = 1500–2000 NH_4^+ = 300-450 TN = 350-500	94	91	74
A²/O	COD = 876–1987 TKN = 84–409	90-97	-	87-98
Activated sludge	TOC = 1009 TN = 254	95.03 (TOC)	-	73.46
Aerobic SBR	COD = 5000 TN = 360	95-96	-	95-97
ABR-AS	TOC = 90.41-1694 TN =161.2-254.7	85.03	-	72.10
MBR	COD = 385–1300 NH_4^+ = 33-51 TN = 90-100	94.1	99.6	93.1

Conventional nitrification and denitrification proceed slowly because of the slow growth of the microorganisms responsible for the conduction of these reactions. In addition, it is difficult to supply the aerobic and anoxic conditions needed for nitrification and denitrification, respectively. To deal with these problems, various types of biological reactors have been tested to enhance conventional nitrification and denitrification. Examples of these efforts include the simultaneous nitrification and denitrification (SNDN), fixation of bacteria on polymeric gel beads in a moving bed biofilm reactor (MBBR) and formation of the biofilm on the surface of rotating disks or other beds in a moving bed biofilm reactor or aeration tank. Unfortunately, these enhanced processes were not effective when applied for the treatment of wastewaters with a high concentration of nitrogen. The main disadvantages of these processes included poor performance because of the low nitrification and denitrification rates, low stability of fixed bacteria and insufficient/unavailable carbon source for denitrification [7, 9 - 14]. However, a number of novel biological nitrogen removal processes were identified in the last decades. Characteristics of these processes along with their performances are described below.

NOVEL BIOLOGICAL PROCESSES FOR NITROGEN REMOVAL
SHARON PROCESS

The SHARON process (single reactor high activity ammonia removal over nitrite) was first raised by Hellinga *et al.* (1998). This process was developed for the treatment of ammonium-rich supernatant of a centrifugal sludge dewatering process. The supernatant of the centrifuged digested sludge contains about 1 gNH$_4^+$-N/L. The SHARON process is more suitable for high ammonium concentration since ammonium conversion is decreased from >97% to about 50% when influent ammonium concentration decreases from 5000 to 250 mgNH$_4^+$-N/L. The SHARON process differs from other biological wastewater treatment processes because of the complete absence of biomass retention in the reaction tank, so that the growth and washout of sludge are in equilibrium. This process is done without any sludge retention in a single aerated reactor with relatively high temperature (35°C) and pH (> 7). The SHARON process is a partial nitrification (nitritation) system, in which half of the influent ammonia nitrogen oxidizes to nitrite (Eq. 12) [7, 9, 11, 12, 15 - 20].

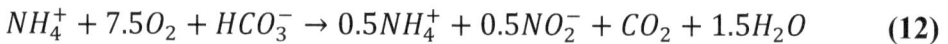

$$NH_4^+ + 7.5O_2 + HCO_3^- \rightarrow 0.5NH_4^+ + 0.5NO_2^- + CO_2 + 1.5H_2O \qquad (12)$$

The process has the following characteristics [7, 9, 11, 12]:

- First studied for biological nitrogen removal over nitrite in wastewaters with high ammonium concentration;
- Because of high temperatures (30–40°C) in the reactor, fast-growing microorganisms (AOB) will predominate. The aerobic hydraulic retention time may behold close to 1 day.
- Due to the high activity of microorganisms in the reactor, the K$_s$ value is rather high. So, effluent substrate concentrations (ammonium) will be several tens of milligram. As the effluent concentration of ammonium is independent from its influent concentration, the removal efficiency increases with increasing the influent concentrations. In the case of digestion supernatant (NH$_4^+$–N >1 g/l), this results in removal efficiencies of over 90%.
- Since at higher temperatures, the NOB grows slower than the AOB, nitrite oxidation can be inhibited (Fig. **1**). Because in this process, SRT is equal to HRT, it is simple to limit SRT, so that ammonium will be oxidized and nitrite will remain.
- Due to the high ammonium concentrations and, subsequently, high nitrification reaction rates, pH control is very important. The bicarbonate in the anaerobic effluent and the denitrification process offset the acidifying effect of the nitrification. Both will supply 50% of the alkalinity requirement. Furthermore,

the CO_2 stripping needs to be sufficient to ensure all of the bicarbonate will be used.

• Due to the high inlet concentrations, heat production is significant. This phenomenon should be considered in the process design.

• As the process operates without biomass retention, there is no concern about the presence of suspended solids in the effluent, *e.g.* due to temporary difficulty in the sludge dewatering.

• Simple single reactor (continuous stirred-tank reactor (CSTR)) system is required.

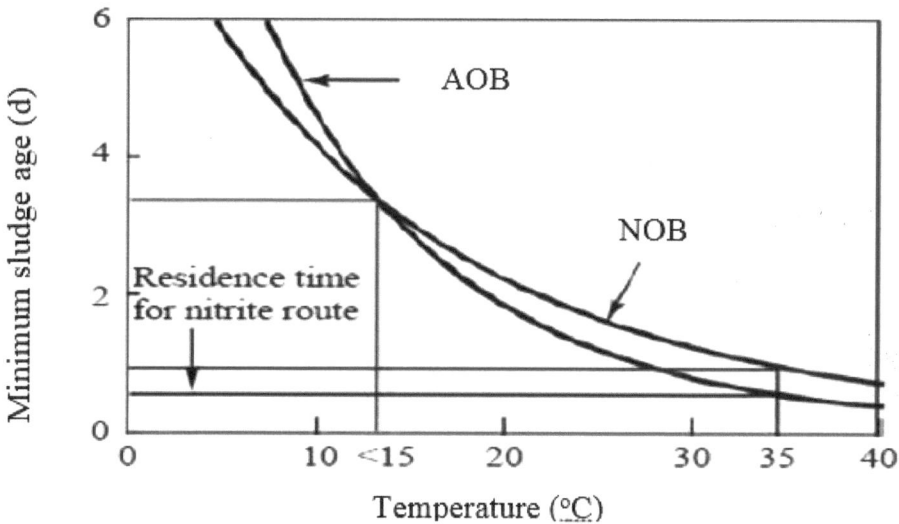

Fig. (1). Minimum residence time for ammonium and nitrite oxidizers at different temperatures (Hellinga *et al.*, 1998) [18].

The biochemistry of AOB in the SHARON process can be defined as follows. In the first stage, ammonium is oxidized to hydroxylamine (NH_2OH) by AMO. In this reaction, O_2 and N_2O_4 act as the electron acceptors for this enzyme (Eqs. (1) and (2)). In the second stage, NH_2OH oxidizes to NO_2^- by the HAO (Eq. (3)). *Nitrosomonas europaea* and *Nitrosomonas eutropha* are the most common AOB. Also, *Nitrosolobus*, *Nitrosopira* and *Nitrosovibrio* can oxidize NH_4^+ to NO_2^-. As mentioned before, although these genera are genetically diverse, they are related to each other in β-Proteobacteria. These genera are resistant to high concentrations of NO_2^-, even higher than 0.5 g NO_2–N/L at pH 7. The physiological characteristics of AOB are presented in Table **3** [7, 9, 11, 12].

Table 3. Kinetics of AOBs.

Coefficients	Quantity
Y (mol/mol C)	0.08
Y (g protein/g NH_4^+–N)	0.1
Aerobic rate (nmol/min/mg of protein)	200–600
Doubling time (day)	0.73
K (h^{-1})	0.04
$K_{s, oxygen}$ (μM)	10–50
$K_{s, ammonium}$ (μM)	5–2600

Effective Factors for SHARON Process

Preventing from the NOB growth is very important for nitritation because they oxidize NO_2^- to NO_3^- and convert nitritation into full nitrification. Several factors including temperature, DO concentration, SRT, pH, aeration pattern, substrate concentration and chemical inhibitors are the selective inhibitor for NOB. NOB require a higher DO concentration than AOB. The DO half-saturation value or affinity constant ($K_{s,O}$) is 62 μM for NOB, whereas this value is very lower for AOB (16 μM). Therefore, at low DO concentrations, AOB dominate NOB, so NO_2^- is accumulated and the nitritation-denitritation can be accrued. Although a low DO concentration (<1.5 mg/L) is favorable for nitritation, it lowers nitritation rates, reduces COD removal efficiencies and interferes with sludge settling. Several DO concentrations have been reported for nitritation (0.3 to 2.5 mg/L). DO concentrations of higher than 2 mg/L could convert nitritation into full nitrification, whereas DO concentrations of lower than 0.5 mg/L could lower nitrification rate. Optimum DO concentrations of 1.0 – 1.5 mg/L have been verified to be suitable for nitritation-denitritation in real municipal wastewater. Table **4** shows the SHARON process responses to different DO concentrations in different systems [7, 9, 11, 12].

Table 4. The SHARON process responses to different DO concentrations in different systems.

Process	DO (mg/L)	Effect
Suspended growth	0.5	Inhibition of nitrite oxidation and its accumulation
	6	Complete nitrification
Activated Sludge	<0.5	Nitrite and ammonium accumulation
	0.7	Up to 67% of the applied NH_4^+ was accumulated as nitrite
	1	80% oxidation of NH_4^+, 80% NO_2^- accumulation
	1.4	oxidation of 99% of NH_4^+, 70% NO_2^- accumulation
	>1.7	Complete nitrification

(Table 4) cont.....

Process	DO (mg/L)	Effect
Biofilm airlift	1.02 1.5 >2.5	Stable and 100% nitrite accumulation 50% of ammonium conversion was accumulated as nitrite Complete Nitrification
Biological aerated filter	2-5	Up to 60% of total ammonia conversion was accumulated as nitrite
Completely Stirred Biofilm	0.5	Complete ammonium removal that 90% of converted ammonium was accumulated as nitrite

Shalini *et al.* (2012) [15]

Nitrification is an acidifying process; therefore, the pH control is important to prevent process inhibition. NOB are particularly sensitive to the variable pH. When pH falls below 6.5, the ammonium oxidation will inhibit because of a pH-dependent equilibrium between the concentrations of NH_3 and NH_4^+. When pH is depressed too low, the NH_3 concentration becomes too low for the sufficient growth of AOB. Although NOB grows faster than AOB at low pH values, the opposite is the case at high pH values. Therefore, a high pH value is preferred for obtaining an effluent that is low at NH_4^+ concentration. In the pH value of above 8, nitrification also inhibits because the high concentration of NH_3 is apparently toxic for NOB. The NH_4^+/NO_2^- ratio in the effluent of the SHARON process can be sensitively influenced by varying the reactor pH between 6.5 and 7.5. Typically, the ratio of HCO_3^-/NH_4^+ ratio for the sludge liquor is between 1-1.1. Consequently, about half of the ammonium in the liquor can be oxidized without any pH control and this phenomenon decreases the alkalinity of water. Alkalinity consumption leads to pH depression and inhibits further nitrification. Based on Arrhenius equation (Eq. (13)), the maximum growth rate (μ_m) of nitrifying bacteria and temperature (at 5–40°C) is defined as follows [7, 9, 11, 12].

$$\mu_{mt} = \mu_{m,20} exp \left[\frac{-E_a(20-t)}{293R(273+t)} \right] \qquad (13)$$

In this equation, μ_{mt} is the maximum specific growth rate (d^{-1}), $\mu_{m,20}$ is the maximum specific growth rate at 20°C (d^{-1}), E_a is the activation energy (kJ/mol) and R is a constant of 8.314 (J/mol.K). Growth rates of AOB and NOB vary with temperature. AOB have a higher maximum specific growth rate (0.801 d^{-1}) than NOB (0.788 d^{-1}) at 20°C, while the specific growth rate of AOB (0.523 d^{-1}) is lower than that of NOB (0.642 d^{-1}) at 15°C. Therefore, NOB dominates AOB at temperatures below 15°C and AOB outcompete NOB at the temperatures of above 20°C. A higher temperature not only enhances the growth of AOB, but can also increase the growth rate differences between AOB and NOB. The SHARON process has been successfully conducted at 35°C, at which AOB become dominant. Because of different bacterial growth rates in SHARON, the microbial community should be selected wherein AOB are retained in the system and NOB is washed out of the system. Conduction of the SHARON process reduces by 25% in oxygen requirement for nitrification. Also, 40% of the external carbon source addition is saved for consequent denitritation. NOB (*e.g.*, *Nitrobacter*) need a shorter retention time than AOB (*e.g.*, *Nitrosomonas*) at the temperatures of below 15°C, while the trend is reversed at the temperatures of above 25°C. Therefore, AOB and NOB can be selectively retained in the reactor by appropriate adjusting SRT in a suspended growth system. SRT is equal to HRT in the SHARON process. With an aerobic HRT below 2 d, ammonia nitrogen can be removed *via* nitrite path in a SHARON process. At an aerobic HRT of approximately 1.5 days, the COD/N ratio clearly shows the metabolic pathways from ammonia to nitrogen gas *via* nitrite. In addition to SRT control, the aeration pattern can be used as an alternative for nitritation control. Aeration time is inversely related to the conduction of partial nitrification because nitritation will be converted into complete nitrification at long aeration times. It has been also reported that NO_2^- could accumulate during a change from anoxic to aerobic conditions. The NO_2^- accumulation prolongs by 2–3 h in the aerobic condition. Intermittent aeration is advantageous for partial nitrification, wherein nitritation can be accomplished by the aeration control strategy, even though it has been reported that the temperature can be decreased from 32°C to 21°C. According to the cell growth rates, AOB are divided into two groups of slow- and fast-growing. Slow-growth AOB, defined as K group, have high affinity to the substrate (low $K_{s,ammonia}$) and are dominant at low ammonia concentrations, whereas fast-growth AOB, defined as R group, have low affinity to the substrate (high $K_{s,ammonia}$) and dominate at high ammonia concentrations. Because ammonia concentrations are normally below 5 mg/L in wastewater treatment processes to meet the discharge requirements, K group may be dominant. R group is dominant in nitritation processes at high ammonia concentrations (>50 mg/L). Several chemical inhibitors suppress NOB and lead to nitritation. Heavy metals such as Cr, Hg, Ag, Ni, Zn, Pb and Cu can enhance nitritation. Also, organic compounds such as phenol, *o*-cresol and aniline show stronger inhibitions on NOB than AOB. Wastewater containing these compounds might inhibit NOB and promote the accumulation of nitrite. Oxidants such as chlorate and ClO_2^- can also inhibit full nitrification. Seawater or saline wastewater containing a high level of ClO_2^- can extend nitritation. High concentrations of free ammonia (FA) and free nitrous acid (HNO_2) are also appropriate for nitrite accumulation. It has been reported that HNO_2-N at the concentrations of $0.2 - 0.22$ mg/L inhibit the NO_2^- oxidation. It is also reported that HNO_2 starts inhibiting the anabolism of *Nitrobacter* at the concentration of 0.011 mg HNO_2^--N/L (0.8 μM) and completely inhibits the biomass synthesis at the concentration of 0.023 mg HNO_2–N/L (1.6 μM). Many laboratory-scale systems have reported achieving stable nitritation-denitritation by the inhibition of free ammonia (FA). NOB are inhibited by NH_3-N in the concentration range of 0.1–1.0 mg/L, while AOB can resist NH_3-N up to 10 – 150 mg/L. However, inhibiting the activities of AOB and NOB by FA is temporary. When the FA concentration is lowered to 0.2 mg/L, the nitrite oxidation by NOB is recovered. It should be considered that the FA concentration is varied by wastewater pH and temperature, which further affects the stability of nitritation [7, 9, 11, 12, 21].

Applicability Aspects of the SHARON Process

By regulating one of the factors described above, persistent nitritation can be obtained. In the economic aspect, DO concentration is the best control factor. Low DO concentration leads to lower aeration cost, but may decrease COD removal efficiency and ruin sludge settling properties. Furthermore, physical and practical conditions should be considered. For example, it is impractical to raise wastewater temperature to facilitate AOB because of the high specific heat of water. It is necessary to consider cost-effectiveness when using control factors such as DO, temperature, pH and inhibitors. Thus far, the most successful reactor for nitritation *via* nitrite has been reported in sequencing batch reactor. The only study for nitritation in a continuous-flow operational mode was conducted with influent NH_4^+-N of more than 50 mg/L. The current challenge is the implementation of persistent nitritation in continuous-flow operation for the treatment of wastewaters with low ammonia concentration (<60 mg/L). The SHARON process can be combined with heterotrophic denitritation (the so-called 'nitrite route', Fig. **2**) as well as the ANAMMOX process [7, 9, 11, 12].

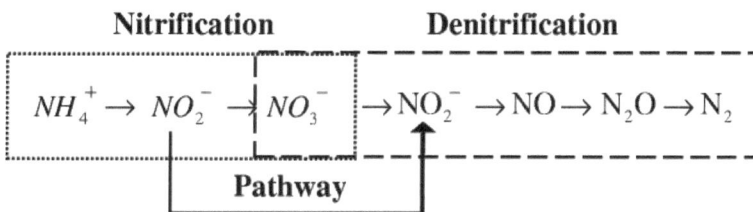

Fig. (**2**). SHARON-denitritation process pathway.

SHARON- HETEROTROPHIC DENITRITATION

When the SHARON process is combined with heterotrophic denitritation, it needs less aeration and the latter denitritation consumes fewer organics because only nitrite and not nitrate has to be reduced to dinitrogen gas (Eqs. (10) and (11)). This is cost-effective for the low C/N ratio wastewaters that need the addition of an external carbon source, such as methanol. In that case, the process also evolves less CO_2 to the air. In this combination, NO_2^- is produced first in the nitritation and, then, it is reduced to N_2 in the following NO_2^- denitritation (Fig. **2**) [7, 9 - 12, 15 - 20].

SHARON-denitritation has the following advantages over traditional nitrification and denitrification:

1. 25% lower oxygen requirement in the aerobic phase that reduces 60% energy consumption in the whole process;
2. Needing for a carbon source as much as 40% lower in the denitritation phase;
3. NO_2^- denitritation rate of 1.5 to 2 times faster than NO_3^- denitrification rate.

It is reported that the SHARON-denitritation is technically feasible and economically favorable, especially when applied for the treatment of wastewater with high ammonia concentration or low C/N ratio. Table **5** summarizes the nitrogen removal from wastewaters with low C/N based on SHARON-denitritation.

Table 5. Nitrogen Removal from the Low C/N Ratio Wastewater through SHARON-denitritation.

Process	C/N (mean)	NO_2^-/NO_x (%)	NH_4^+-N removal (%)	TN*** removal (%)	TN removal rate (kgN/m³.day)
SBR*	3	97	98	98	0.196
SBR	2.1	80	80	74	0.222
SBR	2.8	95.6	95	50	0.08
SBR	2.4	95	95	95	0.238
A/O**	2.9	90	93	63	0.146
A/O	4.4	90	92	87	0.146
A/O	3	95	98	70	

*Sequencing batch reactors (SBR); ** anoxic/oxic (A/O); *** Total Nitrogen (TN)

SHARON is the first full-scale process, in which nitritation/denitritation is achieved with nitrite as the intermediate product. It has been used for treating sludge digestion liquid in the Netherland cities (for example, Rotterdam, Dokhaven, Utrecht, Zwolle and Beverwijk) [7, 9 - 12, 15 - 20].

ANAMMOX PROCESS

Identifying bacteria with capability to oxidize ammonium under anaerobic condition (ANAMMOX bacteria) led to numerous studies for improving ANAMMOX-based processes as alternatives for conventional nitrification/denitrification processes. The ANAMMOX process discovered by Mulder and his co-workers in 1995. In the ANAMMOX process, NH_4^+ anaerobically transforms into dinitrogen (N_2) gas *via* NO_2^- as an electron acceptor. In comparison with conventional processes, this autotrophic process involves a complete conversion of ammonium to nitrogen gas without the addition of organic matter. Hence, it reduces 100% addition or consumption of carbon source and at least needs 50% less oxygen. Because of lower growth rate of the microorganisms responsible for ANAMMOX (than the ones responsible for the nitrification/denitrification process), this process reduces 90% of operational costs related to sludge disposal, besides reducing CO_2 emission. The ANAMMOX is a lithoautotrophic biological nitrogen removal, conducted by a group of *Planctomycete* bacteria. The considerable progress in the molecular biological technique has revealed a great variety of information on the biological diversity of the ANAMMOX bacteria. The ANAMMOX bacteria have a unique physiology. For example, they can oxidize ammonia in the absence of oxygen. ANAMMOX process is highly exergonic and the produced energy can be used by the organisms involved. This process is oxidation of ammonia with nitrite, as the electron acceptor, to produce N_2. Also, the ANAMMOX organisms use CO_2 as the sole carbon source and use nitrite as the electron acceptor. Even though the *Nitrosomonas* species can also oxidize ammonium in the absence as well as presence of oxygen, the "ANAMMOX" is used particularly for nitrogen removal by these *Planctomycetes*-like bacteria [7, 9 - 12, 15 - 20, 22].

The stoichiometry of the ANAMMOX reaction, derived from several basic studies based on mass balance for ANAMMOX enrichment cultures, can be represented by Eq. (14-16).

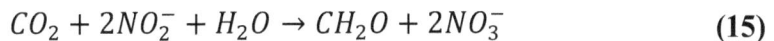

$$NH_4^+ + NO_2^- \rightarrow N_2 + 2H_2O \qquad \text{(14)}$$

$$CO_2 + 2NO_2^- + H_2O \rightarrow CH_2O + 2NO_3^- \qquad \text{(15)}$$

$$NH_4^+ + 1.32NO_2^- + 0.066HCO_3^- + 0.13H^+ \rightarrow 0.066CH_2O_{0.5}N_{0.15} + 0.26NO_3^- + 1.02N_2 + 2.03H_2O \qquad \text{(16)}$$

The main end-product of the ANAMMOX process is N_2, but about 10% of the influent nitrogen is oxidized to nitrate. Based on the overall nitrogen balance, the molar ratio of NH_4^+ to NO_2^- conversion and the ratio of NO_2^- conversion to NO_3^- production is 1:1.31 ± 0.06 and 1:0.22 ± 0.02, respectively. The excess 0.3 moles of NO_2^- is anaerobically oxidized to NO_3^- *i.e.* 1:1.32:0.26 is the ratio of ammonium oxidized, nitrite utilized and nitrate produced, respectively. However, depending on the substrate, operating conditions, and reactor configuration, the NO_2^--N/NH_4^+-N removal in the various ANAMMOX reactors is in the range of 0.5 – 4. ANAMMOX takes place under anoxic conditions, in which ammonia acts as electron donor and nitrite, nitrate, iron III, sulfate, and bicarbonate are the different possible electron acceptors. According to the nature of microorganisms involved in the process, nitrite was shown to be the most common electron acceptor. Recent studies disclosed that ANAMMOX bacteria may not be strictly chemolithotrophic. In addition to ammonium, the ANAMMOX bacteria can use other electron donors such as iron II, and a variety of organic compounds including carboxylic acids (formate, acetate, propionate, methylamines). The characteristics of the ANAMMOX reactions are very considerable and especially appealing. The hydroxylamine and hydrazine, two highly toxic and reactive compounds, are the main ANAMMOX intermediates. The hydrazine differently used as rocket fuel. This compound is not a known intermediate of the other biological nitrogen removal processes. It was observed that hydrazine could be converted by the ANAMMOX culture, using nitrite as the electron acceptor. On the other hand, the ANAMMOX could not grow in the presence of hydrazine, solely. Conclusively, it is notable that the hydrazine acts as an important electron donor in the conversion of nitrite to hydroxylamine in the ANAMMOX process *i.e.* the ANAMMOX is a different process as compared to the other nitrogen conversion processes. The comparisons between ANAMMOX and conventional nitrification is shown in Table **6** [7, 9 - 12, 15 - 20, 23, 24].

The oxidation of both hydrazine and hydroxylamine is catalyzed by hydroxylamine oxidoreductase (HAO) enzyme. This enzyme, also known as hydroxylamine/hydrazine oxidoreductase (HAO), is placed in a membrane-bounded organelle-like organ, existing in the cytoplasm of ANAMMOX organisms. This organelle was defined as "ANAMMOXosome" and believed to be the site of anaerobic ammonium oxidation of ANAMMOX cells. The unique ladderane lipids and hopanoids exist in the membrane of the ANAMMOX bacteria. These lipids are rigid and dense ladders of concatenated cyclobutene rings. The presence of the ladderanes at the center of these organism's membrane makes it extremely impermeable for passive diffusion of chemicals. Ether or ester linked lipids build the ladderane tails. These ladderanes lipids are naturally only present in the ANAMMOX process, and synthetic production of these lipids is very difficult. So, these lipids are used for measurement of the ANAMMOX activity. Two possible pathways for the ANAMMOX process are as follows; in the first pathway, ammonium is oxidized by hydroxylamine to form hydrazine (not *via* NO route) and nitrite is reduced to hydroxylamine, which is coupled to ammonium to produce hydrazine in a cyclic form. Produced electrons, derived from hydrazine, then reduce nitrite for more hydroxylamine and N_2 production (Fig. **3**). Also, nitrate formation could generate reducing equivalents for biomass growth. This pathway was first raised by van de Graaf *et al.*, 1997 derived from ^{15}N studies in a fluidized bed reactor with the dominant species of *Candidatus Brocadia ANAMMOXidans*. The other possible pathway assumes that nitrite is reduced to NO (Fig. **3**) and then a hydrazine forming enzyme, with the uptake of one plus three low energy electrons, combines the produced NO with ammonium to produce hydrazine. Subsequently, a hydrazine–oxidizing enzyme oxidizes hydrazine to N_2. The result of this oxidation is four high energy electrons flowing downhill in the quinine (Q) pool and the H^+ translocating cytochromes bc1 complex, finally generate a PMF that is inside positive. The strongly depleted carbon found experimentally in ladderane lipids of ANAMMOX bacteria with the activity of two enzymes (formate dehydrogenase and carbon monoxide dehydrogenase) in cell-free extracts, and the one-carbon metabolism described in the related *Planctomycetes* are consistent with the acetyl CoA pathway. This mechanism was postulated through the environmental genomics analysis of *Candidatus Kuenenia stuttgartiensis* [7, 9 - 12, 15 - 20].

Table 6. The differences between ANAMMOX and conventional nitrification (Young-Ho, 2006) [12].

Parameter	ANAMMOX	Nitrification
pH range	6.7 – 8.3	Changeable
Temperature (°C)	20 – 43	≤ 42
Free energy	- 367	- 275
Biomass yield (mol/mol C) (g protein/g NH_4^+-N)	0.07 0.07	0.08 0.1
Aerobic rate (nmol/min/mg protein)	0	200 - 600
Anaerobic rate (nmol/min/mg protein)	60	2
Growth rate (h^{-1})	0.003	0.04
Doubling time (days)	10.6	0.73
$K_{S, ammonium}$ (μM)	5	5 – 2600
$K_{S, nitrite}$ (μM)	< 5	n.a.
$K_{S, oxygen}$ (μM)	n.a.	10 – 50
NO_2^- inhibition of NH_4^+ consumption (g NO_2^--N/L)	$K_i = 0.8$, $\alpha = 0.8$	Usually
NO_2^- inhibition of NO_2^- consumption (g NO_2^--N/L)	$K_i = 1$, $\alpha = 0.7$	n.a.
Protein existing in the biomass (g protein/g SS)	0.6	Changeable
Protein density (g protein/L biomass)	50	Changeable

n.a., not applicable; K_S, the affinity constant; K_i, inhibition constant.

The ANAMMOX bacteria, also named as "lithotrophs missing from nature", were found and identified as the new autotrophic members of the major distinct division of bacteria, *Planctomycete*. The ANAMMOX bacterial activity is 25-fold higher than the known aerobic nitrifiers. Until today, *Brocadia ANAMMOXidans, Kuenenia stuttgartiensis, Brocadia fulgida, Scalindua brodae, Scalindua sorokinii, Scalindua wagneri, Scalindua Arabica, Jettenia asiattica,* and *ANAMMOXglobus propionicus* are the discovered ANAMMOX bacteria. The dominant members are *Candidtaus Brocadia ANAMMOXidans* and *Candidatus Kuenenia stuttgartiensis. Nitrosomonas eutropha* and *Nitrosomonas europaea,* two known members of the AOBs, show low ANAMMOX activity in which they use NO_2 as electron acceptor under anoxic conditions [7, 9 - 12, 15 - 20, 22, 23].

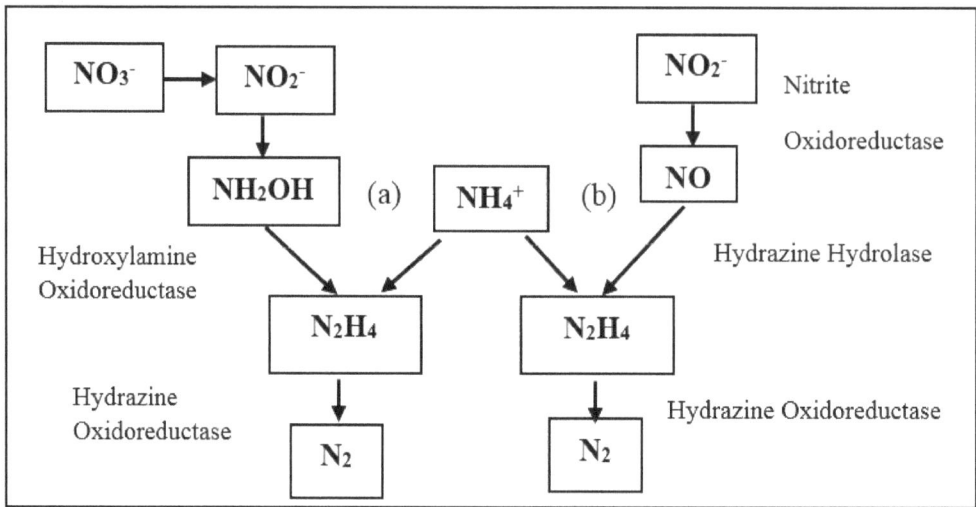

Fig. (3). ANAMMOX process proposed as hydroxylamine route (a) and nitric oxide route (b) (Shalini *et al.* (2012) [15].

The important operational variables of the ANAMMOX process are temperature, pH, C/N ratio, nutrients, and HRT. Two indicators for the ANAMMOX bioreactor are pH and alkalinity. The important variables, mentioned above, were optimized for the ANAMMOX process. Optimum values are as follow; pH in the range of 6.7–8.5, and temperature in the range of 20 - 43 °C (30–37 °C is the most usable range). The ANAMMOX process is suitable for wastewaters with low C/N ratio. At C/N ratio greater than 1, the ANAMMOX bacteria are no longer able to compete with heterotrophic denitrifying bacteria. The HRT of 1 day is optimized for the ANAMMOX process. The activated sludge, nitrification sludge, anaerobic digestion sludge, denitrification sludge (start-up: 105, 100 d, nitrogen removal rate 2.090, 0.609 kgN/m³.d, respectively), and up-flow anaerobic sludge blanket (start-up: 150 d, 0.090 kgN/m³.d) can be used as the seed biomass in the startup of the ANAMMOX reactors. Nutrient requirements should be supplied for the enrichment of the ANAMMOX bacteria. Both macro nutrients (C, N, and P) and microelements (Fe, Co, Ni, Zn, Cu, Mn, and Mo) should be supplied at appropriate values. One of the simplest methods for identifying the ANAMMOX activity is the conduction batch experiments in 100 mL serum bottles with the same amounts of ammonium and nitrite. Also, the presence of ANAMMOX in the reactors can be identified using the following methods [7, 9 - 12, 15 - 20, 22, 23, 25, 26]:

- ANAMMOX activity takes place only with sufficiently high concentrations (at least 10^{10} microorganism/L) of the ANAMMOX cells, and for identifying the organisms responsible for the ANAMMOX, the cell separation *via* density gradient centrifugation can be used.
- Micro-auto-radiography and fluorescent *in-situ* hybridization (FISH).
- PCR using 16SrDNA sequences and 16 rRNA genes.
- Using lipid bio-indicators.
- Measuring the depletion or enrichment of naturally occurring stable isotopes (C^{13} and N^{15}).
- Using Confocal Raman microscopy (CRM) – in this method the distribution of different microorganisms and other substances inside physiological intact microbial communities is determined by a non-invasive technique. FISH-MAR and ISR probing are advanced methods that are used for measuring a single cell activity and growth. Raman-FISH combines stable – isotope Raman spectroscopy and FISH for the single cell analysis to identify and function the ANAMMOX bacteria without pretreating the samples just by its Raman vibrational signature.
- Sensitive detection of the ANAMMOX activity in the reactors using biosensors, which were used for online monitoring.
- Detection of the ANAMMOX activity by ^{15}N labeled ammonium and nitrite
- It was observed that with the increment of nitrogen removal rate, the color alters from khaki to brownish and red.

To identify whether the ANAMMOX bacteria are present in the reactors or absent, these methods can be used, and based on this identification, further enrichment studies can be conducted easily. The ANAMMOX bacteria has a low growth rate (Doubling time 11 days – whereas this value is 1–5 days for aerobic nitrification, conducted by slow-growing autotrophic bacteria) and synthesis yield (0.11 g VSS/g NH_4–N), so the ANAMMOX reactors should have both efficient biomass retention and benefit of low amount of sludge production. Both laboratory and full-scale ANAMMOX process were studied in different reactors. The presence of the ANAMMOX bacteria and ANAMMOX-like activity were observed in the studied reactors. All the studied reactors had good ammonia nitrogen removal. To know more about the process and the bacteria responsible for ammonia transformation, SBR was used more than the other reactors. This reactor enriched a pure culture. To hold the ANAMMOX organisms in the reactor, the sludge settling should be controlled. Although SBR shows more resistance to the hydraulic shock, it is sensitive to substrate shock, conversely, the UASB reactor is the most tolerable to substrate shock. SBR has a homogeneous distribution of substrates, products and biomass aggregates over the reactor, an efficient biomass retention (90%), the experience of reliable operation for more than 1 year and stable conditions under substrate-limiting conditions. However, these advantages can lead to a high degree of enrichment and making make the scale-up easier with contaminants being washed out. The different inhibition studies on the ANAMMOX process showed the way for removal of inhibitory compounds for good enrichment of the ANAMMOX bacteria in the reactors. The results of these studies detailed below [7, 9 - 12, 15 - 20, 23, 24]:

When oxygen and even low organic carbon expose to the enrichment culture, they can completely inhibit the ANAMMOX process. Oxygen, even at concentrations as low as 0.5% or 0.06 mg/L, can inhibit the process, but this inhibition is reversible. It was observed that organic carbon concentration < 300 mg/L can inactivate the ANAMMOX organisms. Co-existence of organic carbon and high concentration of nitrite has a negative effect on both the ANAMMOX bacteria and the heterotrophic denitrifiers. Enough bicarbonate, as the carbon source, can increase the efficiency of the ANAMMOX process in the case of nitrogen removal (nitrogen removal rate significantly rose from 5.2 to 11.8 kgN/m^3. day during 32 days) in an up-flow cylindrical reactor. Also, organic carbon sources, such as propionate and acetate, can be used by the ANAMMOX bacteria to convert nitrite and nitrate but acetate concentrations of 50 mM 70% inhibited the ANAMMOX process. This inhibition concentration is lower for methanol and ethanol, whereas concentrations below 1 mM can inhibit ANAMMOX process. Presence of sulfide at concentrations greater than 1 mM often inhibits the ANAMMOX bacteria. Inhibitory concentration of nitrite is varied, as nitrite concentration in the range of 70–920 mg/L were reported to be detrimental to the ANAMMOX process. To overcome the nitrite inhibition, adding trace amounts of either of the ANAMMOX intermediates (>1.4 mg N/L for hydrazine, >0.7 mg N/L for Hydroxylamine) is helpful. Free ammonia at high concentrations can inhibit the ANAMMOX reaction. The ANAMMOX can tolerate ammonium or its intermediate nitrate up to concentrations of at least 1 g N/L. The affinity constants (K_s) for ammonium and nitrite are lower than 0.1 mgN/L. Phosphate and Acetylene show strong inhibition on ANAMMOX activity. Phosphate concentrations > 180 mg/L inhibit ANAMMOX activity. Proton permeation and the loss of intermediates is detrimental for the ANAMMOX bacteria. To enrich the ANAMMOX bacteria, a large range of bioreactors has been used, and all reactors met the mentioned criteria and enrichment was successful, in an enrichment time of 50 to 200 days, all bioreactors exhibited ANAMMOX activity. In addition, biofilm carriers like Kaldnes rings, glass beads, and non-woven carriers showed to be good alternatives to ensure that ANAMMOX biomass can retain in the used reactors. The ANAMMOX bacteria can grow simultaneously with other kinds of bacteria, like heterotrophs and denitrifying bacteria, and they are less competitive. However, under appropriate conditions, both denitrifying and the ANAMMOX organisms may live and function simultaneously. The simultaneous growth of denitrifiers and the ANAMMOX is technically feasible and economically favorable, particularly for the treatment of wastewater that comprises both ammonium and organic carbon [7, 9 - 12, 15 - 20, 23, 25, 27].

SHARON-ANAMMOX PROCESS IN-SERIES

To gain nitrogen removal *via* ANAMMOX process, nitrite must be present along with ammonium, so an alternative is partial pre-oxidation of the initial ammonium to nitrite (55–60% of ammonium) before feeding into the ANAMMOX process. For this purpose, a nitritation process, such as a SHARON process, must be located before the ANAMMOX process. This combination will change the operational mode of the SHARON process to nitritation (conversion of only 55–60% of ammonium to nitrite) without heterotrophic denitritation, preparing adequate influent condition for the ANAMMOX process. The overall SHARON-ANAMMOX process is described in Eq. (17) [7, 9 - 12, 15 - 20, 23, 25 - 29].

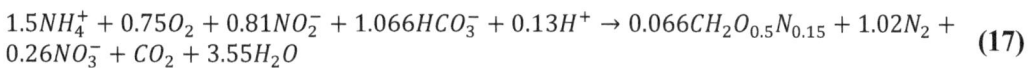

$$1.5NH_4^+ + 0.75O_2 + 0.81NO_2^- + 1.066HCO_3^- + 0.13H^+ \rightarrow 0.066CH_2O_{0.5}N_{0.15} + 1.02N_2 + 0.26NO_3^- + CO_2 + 3.55H_2O \quad (17)$$

This combination has many advantages, including 60% less oxygen requirement, no need to add external carbon source, decreasing sludge production and N_2O emission compared to the conventional nitrification-denitrification process. Comparison of the conventional nitrification-denitrification process with that of the combined SHARON–ANAMMOX process is appealing. The results of this comparison were power consumption: 2.8 and 1.0 kg Wh/kgN, methanol: 3.0 and 0 kg/kgN, sludge production 1.0 and 0.1–0.5 kgVSS/kgN, CO_2 emission >4.7 and 0.7 and operational costs 3.0–5.0 and 1.0–2.0 Euro/kgN, for the conventional nitrification-denitrification and SHARON–ANAMMOX process, respectively. Following these findings, the overall operational costs in treating ammonium-rich wastewater with the SHARON–ANAMMOX process can be saved, compared to that of the conventional processes. As the effectiveness of the SHARON–ANAMMOX system depends on high cell density and excellent biomass retention, a fixed film or immobilized biomass reactor configuration is very helpful. These reactor configurations (fixed or moving bed) increase the concentration of active biomass and subsequently the tolerance to potentially inhibitory operating conditions. Combined processes divided into two general classes, according to the arrangement of two processes. These two classes are one-reactor and two-reactor systems. In one-reactor system, the SHARON and

ANAMMOX processes occur simultaneously in a single reactor (CANON process) and in two-reactor system, the ANAMMOX process takes place next to the SHARON process. It is believed that the one-reactor systems are suitable but more complex than the two reactor systems [7, 9 - 12, 15 - 20, 23, 25, 27 - 29].

CANON PROCESS (SHARON-ANAMMOX IN SINGLE REACTOR)

The CANON process is the combination of partial nitritation and ANAMMOX in a single reactor. In this process, two sequential reactions (Eqs. 18-20) take place in a single aerated reactor. Two groups of bacteria that cooperate in the whole process are *Nitrosomonas*-like aerobic microorganisms and *Planctomycete*-like anaerobic bacteria. In this process, the nitrifiers oxidize ammonia to nitrite aerobically and create anoxic conditions needed for the ANAMMOX bacteria. The operational characteristics of the CANON process that are very sensitive and need to be controlled appropriately are dissolved oxygen, nitrogen-surface load, biofilm thickness, and temperature, *etc.* [7, 9 - 12, 15 - 20, 23, 25, 27, 30, 31].

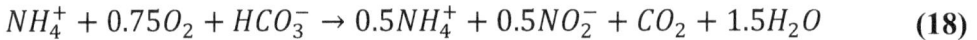

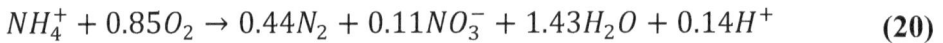

$$NH_4^+ + 0.75O_2 + HCO_3^- \rightarrow 0.5NH_4^+ + 0.5NO_2^- + CO_2 + 1.5H_2O \qquad (18)$$

$$NH_4^+ + 1.32NO_2^- + 0.066HCO_3^- + 0.13H^+ \rightarrow 0.066CH_2O_{0.5}N_{0.1} + 1.02N_2 + 0.26NO_3^- + 2.03H_2O \qquad (19)$$

$$NH_4^+ + 0.85O_2 \rightarrow 0.44N_2 + 0.11NO_3^- + 1.43H_2O + 0.14H^+ \qquad (20)$$

It is proved that the co-existence of aerobic and anaerobic ammonium oxidizing bacteria is feasible under oxygen-limited conditions. As mentioned before, the ANAMMOX bacteria activity could reversely inhibited by low concentrations of oxygen (0.5% air saturation), so providing an environment with the oxygen content below 0.5% air saturation is adequate for a stable co-culture of *Nitrosomonas*-like aerobic microorganisms and *Planctomycete*-like anaerobic bacteria. It was observed that by adding nitrogen oxides (such as NO_2^-) to a mixed culture of *Brocadia ANAMMOXidans* and *Nitrosomonas*, the anaerobic ammonia oxidation occurred with specific rates up to 5.5 mmol NH_4^+/g protein/h for *B. ANAMMOXidans* and 1.5 mmolNH_4^+/g protein/h for *Nitrosomonas*. Also, NO_2^- concentration of 100 mg/L didn't inhibit the ammonia oxidizing activity of *B. ANAMMOXidans*. A model for the competition and cooperation between these two bacteria was derived from experimental results (Fig. **4**).

The CANON process like the SHARON–ANAMMOX process (As mentioned before), needs less oxygen, has lower alkalinity consumption, does not need the addition of external carbon source, and produce less nitrite, nitrate, and negligible amounts of sludge. Finally, this process produces lower unwanted intermediates like NO and N_2O. Quantitatively, NO and N_2O emissions from the SHARON reactor are 0.2% and 1.7% of the nitrogen load, respectively. Also, NO and N_2O emissions from the ANAMMOX reactor are 0.003% and 0.6% of the nitrogen load, respectively. Although the CANON process is more complex, it requires lower investment and operational costs than the other novel processes. Finally, significant savings in terms of energy and resources for treating the ammonium-rich wastewaters by the SHARON-ANAMMOX combination processes can be expected because of the numerous benefits counted for these processes [24, 30 - 34].

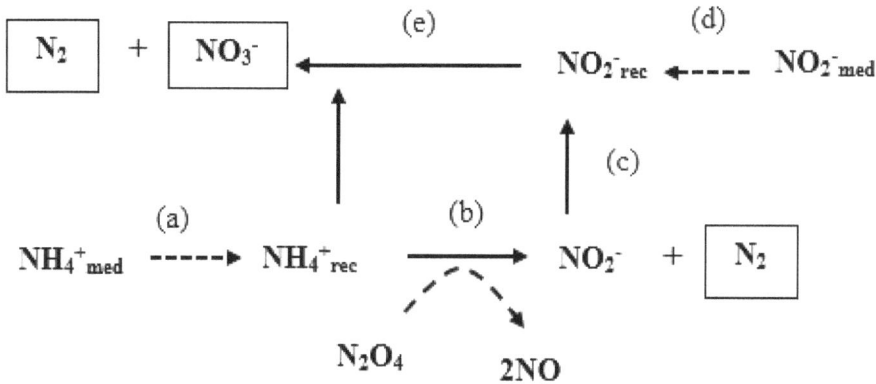

Fig. (4). The proposed model for the interaction between *Brocadia ANAMMOXidans* and *Nitrosomonas* sp. under anoxic conditions. (a) Ammonia addition (fresh medium); (b) oxidation of ammonia by *Nitrosomonas*; (c) oxidation of ammonia to nitrite in the reactor; (d) nitrite addition (fresh medium); (e) ANAMMOX process; med, medium; rec, reactor. enclosed boxes contain end-products (Ahn *et al.* (2006)) [12].

OXYGEN-LIMITED AUTOTROPHIC NITRIFICATION/DENITRIFI-CATION (OLAND) PROCESS

OLAND is a single reactor configuration of SHARON-ANAMMOX process. In this process, in the outer aerobic zones of the biomass, AOB oxidize about half of the ammonium to nitrite (partial nitritation), while in the inner anaerobic zones the ANAMMOX bacteria subsequently convert the residual ammonium by nitrite, as electron acceptor, to mainly N_2 (89%) and some NO_3^- (11%). Oxygen, a key factor in balancing the microbial activities, with ratio of 1.8 g O_2/gN is needed to achieve sufficient ammonium oxidation, while prevents excess nitrite production by AOB. Additionally, to suppress excess nitrate production by NOB, sufficiently low levels of dissolved oxygen (DO) (*e.g.* 0.3 mgO_2/L) are required (Fig. **5**) [7, 9, 11, 12, 35, 36].

Low volumetric carbon and nitrogen loading rates of conventional activated sludge (CAS) systems for wastewater treatment (around 1 gCOD/L.d and 0.08

gN/L.d) and high energy consumption in these systems imposed continuous efforts for replacement of these systems. On the other hand, about 60–70% of the total energy required for a wastewater treatment plant is consumed by aeration to remove organic carbon and nitrogen. However, by an enhanced primary settling to increase physicochemical sludge production, OLAND process can be used for nitrogen removal from supernatant produced during digestion of primary and secondary sludge. Applying this alternative can decrease 25% of aeration requirements of the CAS step. The OLAND process can be applied for low to high-strength nitrogen wastewaters (30-1000 mgN/L), such as concentrated domestic wastewater, landfill leachate, supernatants of sewage sludge digestion, specific industrial effluents, and digested black water, with a range of high hydraulic residence times (HRTs = 0.04 – 1.33 d) (Table 7) [7, 9, 11, 12, 35, 36].

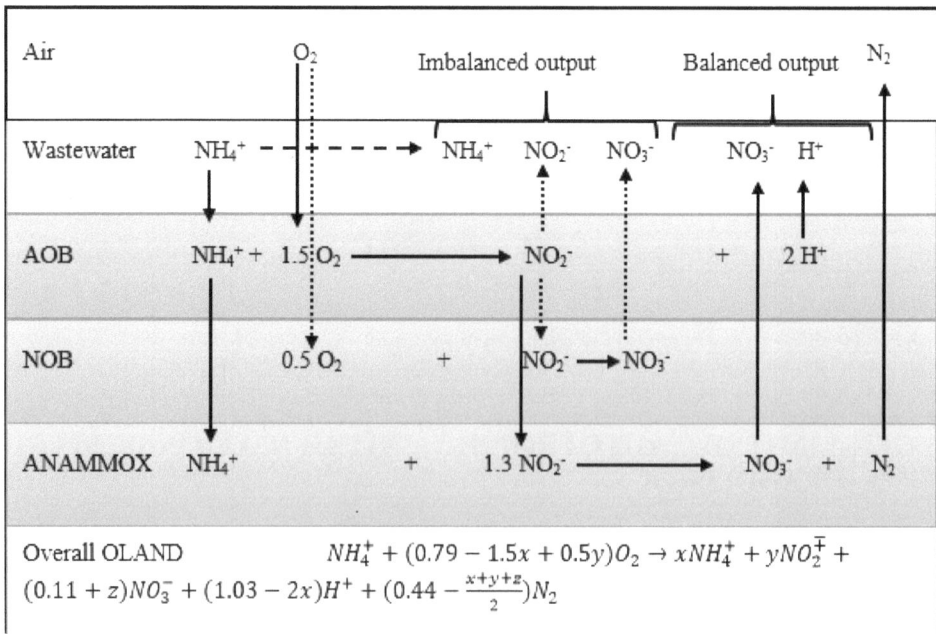

Fig. (5). The proposed pathway for the OLAND process (De Clippeleir *et al.* (2011)) [35].

Table 7. Application of OLAND process for nitrogen removal.

Wastewater	Reactor	NH_4^+ concentration (mgN/L)	N removal rate (gN/L.day)	HRT (d)
Digested black water	RBC	1023	0.71	1.33
Sewage sludge digestate	SBR	800	0.67	0.93

(Table 7) cont.....

Wastewater	Reactor	NH$_4$$^+$ concentration (mgN/L)	N removal rate (gN/L.day)	HRT (d)
Sewage sludge digestate	Gas-lift	650	0.51	1.2
Industrial digestate	RBC	300	1.17	0.18
Landfill leachate	RBC	209	0.38	0.55
Landfill leachate	RBC	250	0.41	0.51
Sewage-like nitrogen concentrations	RBC	66	0.44	0.08
Sewage-like nitrogen concentrations	RBC	31	0.38	0.04

De Clippeleir *et al.* (2011) [35].

The overall reactions take place in the OLAND process are shown in Eqs. (21-23).

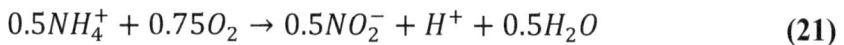

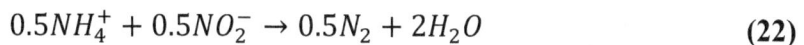

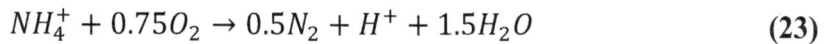

$$0.5NH_4^+ + 0.75O_2 \rightarrow 0.5NO_2^- + H^+ + 0.5H_2O \qquad (21)$$

$$0.5NH_4^+ + 0.5NO_2^- \rightarrow 0.5N_2 + 2H_2O \qquad (22)$$

$$NH_4^+ + 0.75O_2 \rightarrow 0.5N_2 + H^+ + 1.5H_2O \qquad (23)$$

The CANON process is different from that of the OLAND process, as the main difference between these processes is that in the OLAND process the denitrification is done by the conventional aerobic nitrifiers, whereas in the CANON process the ANAMMOX bacteria are present. On the other hand, these processes may be based on the CANON concept of co-existence of aerobic nitrifiers and the ANAMMOX bacteria under oxygen limitation. These processes have been studied in both pilot and full-scale for treatment of ammonium-rich wastewaters. The mechanisms of these processes have many unknown aspects, and the nitrogen loading and removal are totally low [7, 9, 11, 12, 34 - 36].

OTHER PROCESSES

NO$_X$ Process

By adding trace amounts of nitrogen oxides (ratio of ammonium to nitrogen oxides is 1000–5000:1) into wastewater for controlling and stimulation of denitrification activity of *Nitrosomonas*-like bacteria, the NO$_X$ process can be obtained. This process operates at aerobic conditions and the supplemented NO$_X$ (NO/NO$_2$) role is regulating and inducing the nitrification and denitrification activity of *Nitrosomonas*-like bacteria. This process is a class of combined nitrification/denitrification and involved reactions are shown in Eqs. (24-26). The

[H] reactant represents the reducing equivalents (*e.g.* divided from an external carbon source). The NO_x process was applied in laboratory and pilot-scale nitrification plants. In a pilot-scale plant treating ammonium-rich wastewater (about 2 $kgNH_4^+$-N/m^3) with the addition of 200 mg/L NO_2^-, in average, nitrogen transformation to N_2 gas was about 67%. It was evidenced strongly that AOB were dominant bacteria in nitrogen conversion. Based on operating results, it was observed that denitrification activity of the nitrifying bacteria was very sensitive towards the NO_2^- addition [7, 12]. The [H] shows the reducing equivalents (*e.g.* supplied by an external C-source).

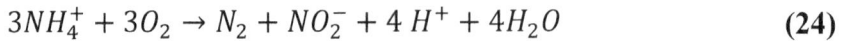

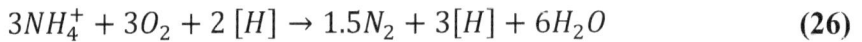

$$3NH_4^+ + 3O_2 \rightarrow N_2 + NO_2^- + 4\,H^+ + 4H_2O \tag{24}$$

$$NO_2^- + H^+ + 3[H] \rightarrow 0.5N_2 + 2H_2O \tag{25}$$

$$3NH_4^+ + 3O_2 + 2\,[H] \rightarrow 1.5N_2 + 3[H] + 6H_2O \tag{26}$$

Denitrifying Ammonia Oxidation (DEAMOX) Process

Nitrite requirement for the ANAMMOX bacteria can be obtained through two ways including nitritation (Eq. 27) and denitratation (Eq. (28)) [37 - 42].

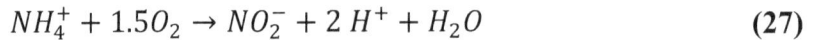

$$NH_4^+ + 1.5O_2 \rightarrow NO_2^- + 2\,H^+ + H_2O \tag{27}$$

$$NO_3^- + 2H^+ + 2e^-(COD) \rightarrow NO_2^- + H_2O \tag{28}$$

The nitritation has been studied in the last decade and a number of processes exploring manipulations of DO concentrations, inhibition/suppression or the temperature-dependent wash-out (*i.e.*, the SHARON process) of NOB have been proposed. The denitratation method for generating nitrite occurred in a system where ANAMMOX was originally found. After extensive development, this approach has been further studied. The result of these studies was the so-called DEAMOX (Denitrifying Ammonia Oxidation) process, where the ANAMMOX reaction is combined with autotrophic denitratation using sulfide as the electron donor [37 - 41].

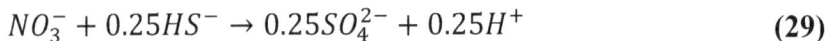

$$NO_3^- + 0.25HS^- \rightarrow 0.25SO_4^{2-} + 0.25H^+ \tag{29}$$

Since the standard application of autotrophic denitratation, using sulfide as an electron donor, is limited to sulfur containing wastewater, the use of heterotrophic denitratation, *e.g.*, using volatile fatty acids (VFA) as a more widespread electron donor group in common wastewater treatment, is more appropriate. So, when acetate (a known VFA) is used as the electron donor, Eq. (29) can be presented as follow [30, 37 - 41]:

$$NO_3^- + 0.25CH_3COO^- \rightarrow 0.5HCO_3^- + 0.25H^+ \tag{30}$$

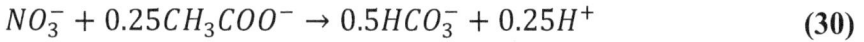

The overall equation (Eq. (14) + Eq. (30)) for heterotrophic denitratation and ANAMMOX is shown in Eq. (31):

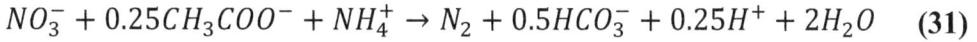

$$NO_3^- + 0.25CH_3COO^- + NH_4^+ \rightarrow N_2 + 0.5HCO_3^- + 0.25H^+ + 2H_2O \tag{31}$$

However, there is a severe competition for nitrite between heterotrophic denitritation and autotrophic ANAMMOX [37 - 42].

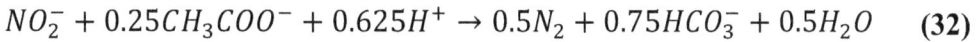

$$NO_2^- + 0.25CH_3COO^- + 0.625H^+ \rightarrow 0.5N_2 + 0.75HCO_3^- + 0.5H_2O \tag{32}$$

There are two types of DEAMOX process, including organics-driven DEAMOX process and sulfide-driven DEAMOX process. Initially, the DEAMOX process was developed in the sulfide-driven modification, then it was extended to the organics-driven one. All the experiments for the sulfide-driven DEAMOX were performed with baker's yeast wastewater in a laboratory scale consisted of three reactors. The UASB reactor can be used as a pretreatment step for generation of required sulfide and ammonia. In the next step, the anaerobic effluent is divided into two streams; firstly, a stream is fed to the nitrifying reactor to generate nitrate and it combines in the effluent with the second stream that directly fed to the DEAMOX reactor. The organics-driven DEAMOX was studied with synthetic wastewater (nitrate + ammonia + VFA) and reject water after thermophilic anaerobic sludge digestion. For synthetic wastewater treatment, a single DEAMOX reactor is used. For nitrogen removal from reject water, the laboratory set-up consisted of two reactors of nitrifying and DEAMOX. In the standard reactor configuration, the nitrifying reactor is located before the DEAMOX reactor. In the reverse reactor order, the nitrifying reactor is located next to the DEAMOX reactor and the nitrifying effluent is recycled into the DEAMOX reactor. The performances of two types of DEAMOX process for different wastewaters are shown in Table **8** [30, 37 - 42].

As seen, almost independently on the nitrogen loading rate (NLR), the sulfide-driven DEANOX process showed high ammonia (> 71%) and total nitrogen (> 84%) conversions. On the contrary, the organics-driven DEAMOX process (treatment of VFA-containing synthetic wastewater) showed low COD removals because of the competition of unwanted heterotrophic denitrifiers. The typical ammonia and total nitrogen removals efficiencies, using the stoichiometric requirements, were obtained to be 40 and 80%, respectively. The experiments on the treatment of real reject water by the organics-driven DEAMOX process (standard reactor sequence, Fig. (**6**) (above)) showed a similar manner and a stable process performance up to NLRs of 1243 mg N/L.d. In this study, removal efficiencies of ammonia, NO_x, and TN were 40, 100 and 66%, respectively. The organics-driven DEAMOX process, with another configuration (reverse reactor sequence, Fig. (**6**) (below)), was successfully tested for real reject water treatment. When the recycle ratio of nitrified effluent to the DEAMOX reactor was 3.45, according to influent flow, the average total nitrogen removal of 84% was observed [30, 37 - 42].

Table 8. Steady-state performance of the sulfide-driven and organic-driven DEAMOX reactor for different wastewater types.

Wastewater	Baker's yeast wastewater[a]	Synthetic wastewater[b]	Reject water[b]
NLR ((mgN/L.d)	318-858	324-1236	361-1243
HRT (days)	0.27-0.64	0.31-1.18	0.32-0.96
Influent NH_4^+-N/NO_3^--N	0.9-1.2	0.99	0.97-1.6
Influent H_2S-COD/ NO_3^--N	1.84-2.22	-	-
Influent COD_{tot}/ NO_3^--N	-	2.2-2.29	4.48-4.83
NH_4^+ removal (%)	71-77	34-42	39-47
NO_x removal (%)	91-96	74-83	93-100
Total N_{inorg} removal (%)	84-87	56-61	66-71
Consumed H_2S-COD/NO_3^--N	1.76-1.94	-	-
Consumed COD_{tot} /NO_3^--N	-	2.65-2.98	3.21-3.5
Consumed NH_4^+/ NO_x^-	0.66-0.69	0.49-0.53	0.46-0.62

a: Sulfide-driven and b: Organic-driven DEAMOX [37 - 41].

Both the developed DEAMOX process and the currently most marketed SHARON–ANAMMOX combination have the same advantages as reagent-less character (even for alkalinity), and a high efficiency (total nitrogen removal >80%) for treatment of strong nitrogenous wastewater with low COD/N ratio. Although nitrogen loading rates in the ANAMMOX reactors are almost two times of the DEAMOX reactor (up to 10 kg N/m^3.day), the DEAMOX process is less complex because conventional units (activated sludge nitrification and UASB-type reactor) are used rather than a highly controlled SHARON reactor for generation of toxic and reactive nitrite. The SHARON reactor may easily intrude the next ANAMMOX reactor, so controlling influent composition for this reactor is required. On the contrary, nitrite accumulation in both DEAMOX reactors is miserable, causing no control on the process. In general, Proteobacteria and *Bacteroidetes* groups are numerically dominant in the DEAMOX reactor and they comprise 32% and 15% of the total 16S rRNA gene library, respectively. The two DEAMOX process modifications (sulfide and organics-driven) were compared, considering their performances as well as the result of competition for nitrite between ANAMMOX and denitritation. It was found that the impact of DEAMOX reaction is higher in the sulfur-laden DEAMOX, because of a lower occurrence of autotrophic denitritation compared to heterotrophic one. However, since the application of sulfur- driven DEAMOX process is limited to sulfur-laden wastewater, the organics-driven modification can also be used as a reliable alternative for strong nitrogenous wastewater with a low COD/N ratio, instead of the conventional nitrification–denitrification process in an economically reasonable way [30, 34, 37 - 42].

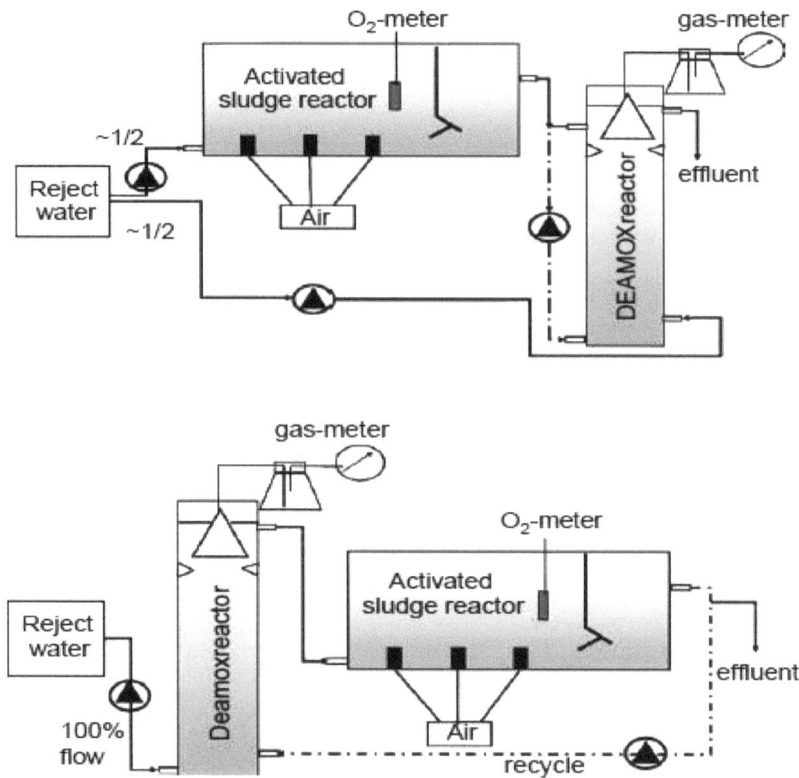

Fig. (6). Experimental set-ups of organics-driven DEAMOX process (standard reactor sequence (above) and reverse reactor sequence (below)) Kalyuzhnyi *et al.* (2009) [37].

Aerobic Deammonification

This process has been described by Hippen *et al.* (1997) for the first time. The nitrogen conversion in the aerobic deammonification process is as follow [11, 43 - 45]:

$$NH_3 + O_2 \rightarrow NH_2OH + H_2O \rightarrow HNO_2 \qquad (33)$$

$$HNO_2 \rightarrow 0.33N_2 + 1.33H_2O + 0.33NO_2^- \qquad (34)$$

$$NH_3 + O_2 \rightarrow 0.33N_2 + 1.33H_2O + 0.33NO_2^- \qquad (35)$$

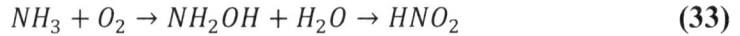

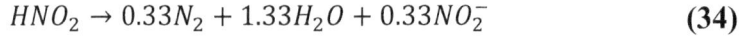

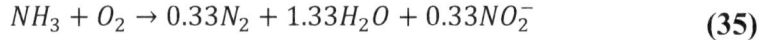

The main difference between this process (similar to OLAND process) and the CANON is that aerobic deammonification and OLAND use denitrification activity of conventional aerobic nitrifiers, whereas the CANON uses the ANAMMOX bacteria. As mentioned in the CANON section, these processes may be based on the CANON concept, in which aerobic nitrifiers and anaerobic ammonia oxidizers cooperate under oxygen limitation. In aerobic deammonification, oxidation of NH_4^+ to N_2 is performed in a single step. For aerobic deammonification process, two models have been proposed. These models are simultaneous nitrification and denitrification model (Fig. **7**, above) and the separated nitrification and denitrification model (Fig. **7**, below). In the first model, aerobic deammonification is conducted by aerobic nitrifiers and the ANAMMOX bacteria (Fig. **6**, above). Aerobic deammonification can be achieved by alternate aerobic/anoxic biofilm reactors. It was observed that in biological rotating contactors, on the surface layer of the biofilm, the *Nitrosomomas* (aerobic AOB) oxidizes ammonia to nitrite using oxygen existing in bulk wastewater, and in inner anoxic layer, the anaerobic AOB convert remained NH_4^+ and produced NO_2^- to N_2. This model has been validated stoichiometrically. *Paracoccus pantotropha* are capable to conduct both anoxic and aerobic denitrifications [11, 43 - 45].

In the separated nitrification and denitrification model, aerobic deammonification is performed by nitrifiers (mainly AOB, such as *Nitrosomomas*) (see Fig. **7**, below). The AOB oxidize NH_4^+ to NO_2^- on the biofilm surface, in the presence of oxygen. Then, the produced NO_2^- is transfused to inner layer of the biofilm and is converted to N_2, as $NADH_2$ acts as the electron donor. In this model, hydroxylamine and $NADH_2$ (produced due to hydroxylamine oxidization) are intermediates of ammonium oxidation. In this model, AOB can only reduce 67% of NO_2^- to N_2. When aerobic deammonification was used for nitrogen removal from municipal wastewater and landfill leachate in both pilot and full-scale, oxidation of NH_4^+ to N_2 occurred. As aerobic deammonification normally takes place in the conventional nitrification, optimization of the process design has not yet been performed and nitrogen loading rates are low (90–250 g N/m^3.d^{-1}). More studies need to be carried out to understand the mechanisms, identify the microbial communities, and improve process control [11, 43, 44, 46].

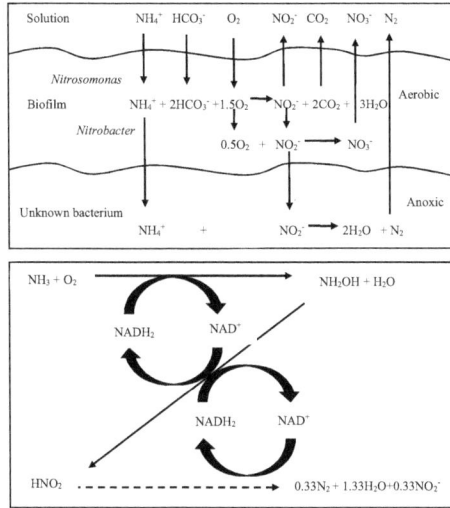

Fig. (7). (Above) Aerobic deammonification based on the simultaneous nitrification and denitrification model and (below) Possible conversion of ammonia to N_2 and NO_2^- (Zhu *et al.* (2008)) [11].

Heterotrophic Nitrification-Aerobic Denitrification

Historically, denitrification was considered as a nearly exclusive or facultative anaerobic or microaerophilic process. With some exceptions, there are no examples that denitrification takes place in an obligate anaerobic condition. On the other hand, conventional nitrification is conducted by autotrophic bacteria. *Arthrobacter* spp., isolated from the natural environment, was the first bacteria capable of heterotrophic nitrification. This species was discovered in 1972. Eleven years later, in 1983, *Thiosphaera pantotropha* (now known as *Paracoccus denitrifican*) was isolated from activated sludge in the wastewater treatment plant. *Thiosphaera pantotropha* showed simultaneous heterotrophic nitrification-aerobic denitrification. Aerobic denitrifiers can use both oxygen and nitrate as final electron acceptors. Nitrogen removal by heterotrophic nitrification and aerobic denitrification has several advantages compared to the conventional nitrogen removal. Firstly, the organisms involved in this process are capable of utilizing organic substrates in the presence of oxygen to achieve simultaneous nitrification and denitrification (SNDN) in a single reactor. In addition, alkalinity produced during the denitrification process can regulate the pH in the reactor, preventing the pH drop caused by nitrification. Furthermore, the diversity of substrates and products of heterotrophic nitrification allow having a variety of organisms and enhance the application scope. The range of oxygen concentrations to conduct aerobic denitrification is wide and differs among various organisms. *Paracoccus* species were the first bacteria capable of complete denitrification, in the presence of high dissolved oxygen (DO) concentrations. Also, *Thiosphaera pantotropha, Alcaligenes faecalis, Microvirgula aerodenitrificans, Citrobacter diversus, Rhodococcus* spp. CPZ24, *Bacillus methylotrophicus* L7, *Acinetobacter calcoaceticus, Magnetospirillum magnetotacticum, Pseudomonas stutzeri* SU2, *Pseudomonas putida, Pseudomonas aeruginosa,* and *Acinetobacter junii YB* were capable of conducting heterotrophic nitrification-aerobic denitrification. Two possible pathways were proposed for heterotrophic nitrification–aerobic denitrification. The difference between them is related to the aerobic denitrification route. In the first proposed pathway, a fully nitrification and denitrification occurs by bacteria and the main genera known to conduct this pathway is *T. pantotropha*. In the other pathway, denitrification takes place through hydroxylamine, not nitrite or nitrate, and the main genera known to conduct this pathway is *A. faecalis*. However, the respiratory pathway has yet to be studied, deeply. It was observed that *Rhodococcus* spp. CPZ24 could remove nitrate and ammonium at rates of 3.1 mg/L.h and 0.93 mg/L.h, respectively. The end-products of denitrification conducted by the strain CPZ24 were reported to be N_2 (almost 99%) and N_2O (about 1%), NO_2 and NO. The rate of NH_4^+-N removal by *bacillus spp.* LY was 0.43 mg/L.h for the initial ammonium concentration of 41.1 mg/L and the C/N ratio of 15. The NH_4^+-N removal rate of 1.15 mg/L.h was obtained for *Pseudomonas alcaligenes* AS-1. The effective parameters in aerobic denitrification are temperature, the carbon/nitrogen (C/N) ratio, pH, dissolved oxygen (DO) concentration, water activity, and the microbial population. The optimal conditions for heterotrophic nitrification by *Bacillus methylotrophicus* strain L7 with sodium succinate as carbon source, obtained from single factor experiments, are C/N = 6, 37 °C, pH = 7–8, 0 g/L NaCl in a wide range of NH_4^+-N concentration (80 - 1000 mg/L). The optimal conditions for aerobic denitrification achieved from orthogonal tests are C/N = 20, DO 4.82 mg/L pH =7–8, and 10 g/L NaCl with nitrite as substrate. Efficient heterotrophic nitrifying–aerobic denitrifying ability was observed by strain L7 with the maximum NH_4^+-N and NO_2^--N removal rate of 2.14 and 0.24 mg/L.h, respectively. It was reported that strain L7 could oxidize ammonium to nitrite, whereas nitrate was not produced. This phenomenon indicates conduction of heterotrophic nitrification by strain L7. The dominant product of the heterotrophic nitrification process was nitrite. The amount of hydroxylamine, an important intermediate in nitrification, was not significantly increased when strain L7 was cultivated in heterotrophic nitrification medium (HNM). It may be because of the instability of hydroxylamine and fast conversion to its intermediates such as nitrite. In the presence of nitrate as the sole nitrogen source, strain L7 reduced nitrate to nitrite indicating that this strain could denitrify nitrate. With the increasing of C/N ratio, the nitrification rate directly increased by *Acinetobacter junii* YB. At C/N ratios of 2, 5, 10 and 15, the maximum NH_4^+-N removal rates of 4.04, 7.25, and 8.98, 10.09 mg /L.h were attained, respectively. Also, at C/N ratios of 2, 5, 10 and 15 within 24 h, TN removal efficiencies of 53.1%, 82.12%, 98.7%, and 99.1% were observed, respectively. The preferred conditions for heterotrophic nitrification by *Acinetobacter junii* YB were C/N = 15, pH 7.5, and 37 °C when succinate was used as the carbon source. Also, *Acinetobacter junii* YB can conduct heterotrophic nitrification–aerobic denitrification in a broad range of NH_4^+-N concentration (50 to 1000 mg/L). From these findings, it can be concluded that heterotrophic nitrification–aerobic denitrification process is compatible with high C/N ratio and can be used for nitrogen removal from raw wastewater with high loads of both nitrogen and organic carbon contaminations [7, 9, 11, 12, 47 - 55].

COMPARING CONVENTIONAL AND NOVEL BIOLOGICAL NITROGEN REMOVAL PROCESSES

Comparisons between conventional and novel biological nitrogen removal is presented in Table **9**. As can be observed, the novel autotrophic processes, including the SHARON, the ANAMMOX, combined SHARON-ANAMMOX, CANON and OLAND processes, consume less energy, need less alkalinity addition and oxygen preparing, produce less nitrite and nitrate, do not produce undesirable by-products like N_2O, do not need addition of chemicals like external carbon source and produce less sludge than the conventional biological nitrogen removal processes. Considering these advantages, it is capable to substantially save energy and resources, especially for ammonium-rich wastewater treatment.

Table 9. Comparison between conventional and novel biological nitrogen removal.

System	Nitrification-Denitrification	OLAND	CANON	SHARON	ANAMMOX
Number of Reactor	2	1	1	1	1
Feed	Ammonium rich wastewater	Ammonium rich wastewater	Ammonium rich wastewater	Ammonium rich wastewater	Ammonium nitrite mixture
Products	N_2, NO_2^-, NO_3^-	N_2, NH_4^+	N_2, NO_3^-	NO_2^-, NH_4^+	N_2, NO_3^-
Conditions	Aerobic, anoxic	Aerobic, anoxic	Oxygen limited	Aerobic	Anoxic
Oxygen requirements	High	Low	Low	Low	None
pH control	Yes	-	None	None	None
Biomass retention	None	Yes	Yes	None	Yes
COD requirement	Yes	None	None	None	None
Alkalinity consumption ($gCaCO_3$/gN)	7.07-3.57	3.6	3.68	-	-

(Table 9) cont.....

System	Nitrification-Denitrification	OLAND	CANON	SHARON	ANAMMOX
Alkalinity consumption $(gCaCO_3/gN)$	7.07-3.57	3.6	3.68	-	-
Sludge production	High	Low	Low	Low	Low
Reactor capacity $(kgN/m^3.day)$	0.05–4	1	1-3	1	6-12
NH_4^+ loading $(kgN/m^3 reactor.d)$	2-8	0.1	2-3	0.5-1.5	10-20
Aerobic ammonia oxidizers (AMO)	Many	Unknown	*N. eutropha*	*N. eutropha*	Absent
Aerobic nitrite oxidizers	Many	Unknown	Absent	Absent	Absent
Anaerobic ammonium oxidizers	Absent	Unknown	*Brocadia ANAMMOXidans/Kunenia stuttgartiensis*	Absent	*B.ANAMMOXidans K.stuttgartiensis*
Overall Bacterial population	Nitrifiers + heterotrophs	Nitrosomonas species	AMO + Planctomycetes	AMO	Planctomycetes
Biofilms or suspension	Biofilms/suspension	Biofilms	Biofilms	Suspension	Biofilms
N-removal efficiency	95%	85%	90%	90%	90%
Process complexity	Separate aerobic and anoxic compartments, Methanol dosing	Aeration needs to be tuned to ammonia loading	Aeration needs to be tuned to ammonia loading	-	Preceding partial nitrification needed
Application status	Established	Laboratory studies	Laboratory studies	Full-scale plants	Full-scale plants
Investment costs	Medium	Medium	Medium	Medium	Low
Operational costs	high	Unknown	Low	Low	Very low

CONCLUDING REMARKS

As mentioned above, all the novel processes, including the SHARON, the ANAMMOX, combined SHARON-ANAMMOX, CANON and OLAND processes, are completely autotrophic, in which ammonium oxidizes to nitrite. Application of these processes to the mainstream food industries wastewater is not possible. If the organic contaminants of food wastewater are removed completely before these processes using anaerobic digestion or high load conventional activated sludge (in which carbonaceous organics are transformed into cells), it is economical to remove nitrogen compounds *via* these processes. Also, if partial removal of organic matter occurs first, the combination of SHARON-heterotrophic denitritation process is feasible. The anaerobic treated food wastewater could offer a critical benefit for SHARON-heterotrophic denitritation process because high alkalinity existing in the anaerobic effluent reduces the need for the addition of external alkalinity source and also could decrease C/N ratio of wastewater to treat *via* SHARON-heterotrophic denitritation process. A study was conducted on the removal of organics and nitrogen from food wastewater using combined thermophilic two-phase anaerobic digestion and SHARON-heterotrophic denitrification process. It was observed that the applied combined system was able to use of about 99% of 121950 ± 23197 mg/L COD and 88% of 2595 ± 589 mg N/L. The SHARON-heterotrophic denitrification process alone showed the average TN removal efficiency of about 74% at very low C/N (TCOD/TN ratio of 2). Using the combined system, organic removal efficiency was improved; also, valuable energy in terms of biogas was obtained. However, nitrogen was converted at a low C/N ratio without the addition of external carbon and alkalinity sources. Due to the anaerobic degradation of proteinaceous compounds to ammonia, the ammonium concentration was increased in the effluent of the two-phase anaerobic digestion reactor. In particular, the ammonium concentration in the anaerobic digestion reactor effluent was much higher than that of the pre-fermentation reactor. The highest ammonium concentration was 2380 mgN/L in the anaerobic digestion reactor, showing that most of the ammonium was produced in the anaerobic digestion reactor. The inhibition concentration of NH_4^+-N for anaerobic digestion is about 3000 mg/L, much higher than the generated concentration. The proper regulation of the operating parameters of SHARON process includes providing the temperature of 35–40°C, pH value in the range of 7.5–8.5 and DO concentration of below 1.5 mg/L, consequently affecting the free ammonia concentration that improves the nitrite accumulation in the system. Since digested food wastewater contains high alkalinity, without the addition of external alkalinity, the pH of the wastewater could be manipulated between 7.0 and 7.5.

Based on Table **1**, in the case of food wastewater, C/N ratio is higher than 4. On the other hand, heterotrophic nitrification-aerobic denitrification process acts at high C/N. So, this process can be used as the mainstream treatment of food industry wastewater, but design parameters should be optimized through the bench and pilot-scale studies. Another alternative for mainstream food industry wastewater treatment is organic-laden DEAMOX process. Like heterotrophic nitrification-aerobic denitrification process, organic-laden DEAMOX process needs to be more accurately studied. There are limit experiences in the pilot and full-scale application of both processes worldwide.

CONSENT FOR PUBLICATION

Not applicable.

REFERENCES

[1]   Wang Y, Huang X, Yuan Q. Nitrogen and carbon removals from food processing wastewater by an anoxic/aerobic membrane bioreactor. Process Biochem 2005; 40(5): 1733-9.
[http://dx.doi.org/10.1016/j.procbio.2004.06.039]

[2]   Johns M. Developments in wastewater treatment in the meat processing industry: A review. Bioresour Technol 1995; 54(3): 203-16.
[http://dx.doi.org/10.1016/0960-8524(95)00140-9]

[3]   Demirel B, Yenigun O, Onay TT. Anaerobic treatment of dairy wastewaters: a review. Process Biochem 2005; 40(8): 2583-95.
[http://dx.doi.org/10.1016/j.procbio.2004.12.015]

[4]   Tenca A, Cusick RD, Schievano A, Oberti R, Logan BE. Evaluation of low cost cathode materials for treatment of industrial and food processing wastewater using microbial electrolysis cells. Int J Hydrogen Energy 2013; 38(4): 1859-65.
[http://dx.doi.org/10.1016/j.ijhydene.2012.11.103]

[5]   Sentürk E, Ince M, Engin GO. Treatment efficiency and VFA composition of a thermophilic anaerobic contact reactor treating food industry wastewater. J Hazard Mater 2010; 176(1-3): 843-8.
[http://dx.doi.org/10.1016/j.jhazmat.2009.11.113] [PMID: 20022175]

[6]   Fongsatitkul P, Wareham DG, Elefsiniotis P, Charoensuk P. Treatment of a slaughterhouse wastewater: effect of internal recycle rate on chemical oxygen demand, total Kjeldahl nitrogen and total phosphorus removal. Environ Technol 2011; 33(15-16): 1755-9.
[http://dx.doi.org/10.1080/09593330.2011.555421] [PMID: 22439562]

[7]   Schmidt I, Sliekers O, Schmid M, *et al.* New concepts of microbial treatment processes for the nitrogen removal in wastewater. FEMS Microbiol Rev 2003; 27(4): 481-92.
[http://dx.doi.org/10.1016/S0168-6445(03)00039-1] [PMID: 14550941]

[8]   Ali M, Okabe S. Anammox-based technologies for nitrogen removal: Advances in process start-up and remaining issues. Chemosphere 2015; 141: 144-53.
[http://dx.doi.org/10.1016/j.chemosphere.2015.06.094] [PMID: 26196404]

[9]   Khin T, Annachhatre AP. Novel microbial nitrogen removal processes. Biotechnol Adv 2004; 22(7): 519-32.
[http://dx.doi.org/10.1016/j.biotechadv.2004.04.003] [PMID: 15262315]

[10]   Dong H, Wang W, Song Z, *et al.* A high-efficiency denitrification bioreactor for the treatment of acrylonitrile wastewater using waterborne polyurethane immobilized activated sludge. Bioresour Technol 2017; 239 (Suppl. C): 472-81.
[http://dx.doi.org/10.1016/j.biortech.2017.05.015] [PMID: 28544987]

[11]   Zhu G, Peng Y, Li B, *et al.* Biological removal of nitrogen from wastewater. Rev Environ Contam Toxicol. Springer 2008; pp. 159-95.

[12]   Ahn Y-H. Sustainable nitrogen elimination biotechnologies: a review. Process Biochem 2006; 41(8): 1709-21.

[http://dx.doi.org/10.1016/j.procbio.2006.03.033]

[13] Henze M. Capabilities of biological nitrogen removal processes from wastewater. Water Sci Technol 1991; 23(4-6): 669-79.
[http://dx.doi.org/10.2166/wst.1991.0517]

[14] Wang LK, Shammas NK, Hung Y-T. Advanced Biological Treatment Processes. New Delhi, India: Springer Humana Press, CBS Publishers 2010; XXII: p. 738.
[http://dx.doi.org/10.1007/978-1-60327-170-7]

[15] Sri Shalini S, Joseph K. Nitrogen management in landfill leachate: application of SHARON, ANAMMOX and combined SHARON-ANAMMOX process. Waste Manag 2012; 32(12): 2385-400.
[http://dx.doi.org/10.1016/j.wasman.2012.06.006] [PMID: 22766438]

[16] Valverde-Pérez B, Mauricio-Iglesias M, Sin G. Systematic design of an optimal control system for the SHARON-ANAMMOX process. J Process Contr 2016; 39: 1-10.
[http://dx.doi.org/10.1016/j.jprocont.2015.12.009]

[17] Van Hulle SW, Vandeweyer HJ, Meesschaert BD, *et al.* Engineering aspects and practical application of autotrophic nitrogen removal from nitrogen rich streams. Chem Eng J 2010; 162(1): 1-20.
[http://dx.doi.org/10.1016/j.cej.2010.05.037]

[18] Hellinga C, Schellen A, Mulder JW. van Loosdrecht Mv, Heijnen J. The SHARON process: an innovative method for nitrogen removal from ammonium-rich waste water. Water Sci Technol 1998; 37(9): 135-42.
[http://dx.doi.org/10.2166/wst.1998.0350]

[19] Mulder JW, van Loosdrecht MC, Hellinga C, van Kempen R. Full-scale application of the SHARON process for treatment of rejection water of digested sludge dewatering. Water Sci Technol 2001; 43(11): 127-34.
[http://dx.doi.org/10.2166/wst.2001.0675] [PMID: 11443954]

[20] Van Dongen LGJM, Jetten MSM, Van Loosdrecht MCM. The combined Sharon/ANAMMOX Process: a sustainable method for N-removal from sludge water.STOWA Report. London: IWA Publishing 2001; p. 63.

[21] Wang L, Zheng P, Abbas G, *et al.* A start-up strategy for high-rate partial nitritation based on DO-HRT control. Process Biochem 2016; 51(1): 95-104.
[http://dx.doi.org/10.1016/j.procbio.2015.11.016]

[22] Zhang L, Zheng P, Tang CJ, Jin RC. Anaerobic ammonium oxidation for treatment of ammonium-rich wastewaters. J Zhejiang Univ Sci B 2008; 9(5): 416-26.
[http://dx.doi.org/10.1631/jzus.B0710590] [PMID: 18500782]

[23] Kuenen JG. Anammox bacteria: from discovery to application. Nat Rev Microbiol 2008; 6(4): 320-6.
[http://dx.doi.org/10.1038/nrmicro1857] [PMID: 18340342]

[24] Sliekers AO, Third KA, Abma W, Kuenen JG, Jetten MS. CANON and Anammox in a gas-lift reactor. FEMS Microbiol Lett 2003; 218(2): 339-44.
[http://dx.doi.org/10.1016/S0378-1097(02)01177-1] [PMID: 12586414]

[25] Terada A, Zhou S, Hosomi M. Presence and detection of anaerobic ammonium-oxidizing (ANAMMOX) bacteria and appraisal of ANAMMOX process for high-strength nitrogenous wastewater treatment: a review. Clean Technol Environ Policy 2011; 13(6): 759-81.
[http://dx.doi.org/10.1007/s10098-011-0355-3]

[26] Jetten MS, Schmid M, Schmidt I, *et al.* Improved nitrogen removal by application of new nitrogen-cycle bacteria. Rev Environ Sci Biotechnol 2002; 1(1): 51-63.
[http://dx.doi.org/10.1023/A:1015191724542]

[27] Chen H, Liu S, Yang F, Xue Y, Wang T. The development of simultaneous partial nitrification, ANAMMOX and denitrification (SNAD) process in a single reactor for nitrogen removal. Bioresour Technol 2009; 100(4): 1548-54.

[http://dx.doi.org/10.1016/j.biortech.2008.09.003] [PMID: 18977657]

[28]   Huang X, Urata K, Wei Q, *et al.* Fast start-up of partial nitritation as pre-treatment for ANAMMOX in membrane bioreactor. Biochem Eng J 2016; 105: 371-8.
[http://dx.doi.org/10.1016/j.bej.2015.10.018]

[29]   Li H, Zhou S, Ma W, *et al.* Long-term performance and microbial ecology of a two-stage PN-ANAMMOX process treating mature landfill leachate. Bioresour Technol 2014; 159: 404-11.
[http://dx.doi.org/10.1016/j.biortech.2014.02.054] [PMID: 24681301]

[30]   Paredes D, Kuschk P, Mbwette T, *et al.* New aspects of microbial nitrogen transformations in the context of wastewater treatment–a review. Eng Life Sci 2007; 7(1): 13-25.
[http://dx.doi.org/10.1002/elsc.200620170]

[31]   Figueroa M, Vázquez-Padín JR, Mosquera-Corral A, Campos JL, Méndez R. Is the CANON reactor an alternative for nitrogen removal from pre-treated swine slurry? Biochem Eng J 2012; 65: 23-9.
[http://dx.doi.org/10.1016/j.bej.2012.03.008]

[32]   Liu T, Li D, Zhang J, Lv Y, Quan X. Effect of temperature on functional bacterial abundance and community structure in CANON process. Biochem Eng J 2016; 105: 306-13.
[http://dx.doi.org/10.1016/j.bej.2015.10.001]

[33]   Lackner S, Gilbert EM, Vlaeminck SE, Joss A, Horn H, van Loosdrecht MC. Full-scale partial nitritation/anammox experiences--an application survey. Water Res 2014; 55: 292-303.
[http://dx.doi.org/10.1016/j.watres.2014.02.032] [PMID: 24631878]

[34]   Sun S-P. Effective biological nitrogen removal treatment processes for domestic wastewaters with low C/N ratios: a review. Environ Eng Sci 2010; 27(2): 111-26.
[http://dx.doi.org/10.1089/ees.2009.0100]

[35]   De Clippeleir H, Yan X, Verstraete W, Vlaeminck SE. OLAND is feasible to treat sewage-like nitrogen concentrations at low hydraulic residence times. Appl Microbiol Biotechnol 2011; 90(4): 1537-45.
[http://dx.doi.org/10.1007/s00253-011-3222-6] [PMID: 21465304]

[36]   Schaubroeck T, Bagchi S, De Clippeleir H, Carballa M, Verstraete W, Vlaeminck SE. Successful hydraulic strategies to start up OLAND sequencing batch reactors at lab scale. Microb Biotechnol 2012; 5(3): 403-14.
[http://dx.doi.org/10.1111/j.1751-7915.2011.00326.x] [PMID: 22236147]

[37]   Kalyuzhnyi S, Gladchenko M. DEAMOX–New microbiological process of nitrogen removal from strong nitrogenous wastewater. Desalination 2009; 248(1-3): 783-93.
[http://dx.doi.org/10.1016/j.desal.2009.02.054]

[38]   Du R, Cao S, Li B, Wang S, Peng Y. Simultaneous domestic wastewater and nitrate sewage treatment by DEnitrifying AMmonium OXidation (DEAMOX) in sequencing batch reactor. Chemosphere 2017; 174: 399-407.
[http://dx.doi.org/10.1016/j.chemosphere.2017.02.013] [PMID: 28187386]

[39]   Cao S, Peng Y, Du R, Wang S. Feasibility of enhancing the DEnitrifying AMmonium OXidation (DEAMOX) process for nitrogen removal by seeding partial denitrification sludge. Chemosphere 2016; 148: 403-7.
[http://dx.doi.org/10.1016/j.chemosphere.2015.09.062] [PMID: 26829308]

[40]   Ma B, Wang S, Cao S, *et al.* Biological nitrogen removal from sewage *via* anammox: Recent advances. Bioresour Technol 2016; 200: 981-90.
[http://dx.doi.org/10.1016/j.biortech.2015.10.074] [PMID: 26586538]

[41]   Nozhevnikova A, Litti YV, Nekrasova V, *et al.* Anaerobic ammonium oxidation (ANAMMOX) in immobilized activated sludge biofilms during the treatment of weak wastewater. Microbiol 2012; 81(1): 25-34.
[http://dx.doi.org/10.1134/S0026261712010110]

[42]   Masłoń A, Tomaszek JA. Anaerobic ammonium nitrogen oxidation in Deamox process. Environ Prot Eng 2009; 35(2): 123-30.

[43]   Kalyuzhnyi S, Gladchenko M, Mulder A, Versprille B. DEAMOX--new biological nitrogen removal process based on anaerobic ammonia oxidation coupled to sulphide-driven conversion of nitrate into nitrite. Water Res 2006; 40(19): 3637-45.
[http://dx.doi.org/10.1016/j.watres.2006.06.010] [PMID: 16893559]

[44]   Hippen A, Rosenwinkel K-H, Baumgarten G, Seyfried CF. Aerobic deammonification: a new experience in the treatment of wastewaters. Water Sci Technol 1997; 35(10): 111-20.
[http://dx.doi.org/10.2166/wst.1997.0371]

[45]   Yi Y-S, Kim S, An S, Choi SI, Choi E, Yun Z. Gas analysis reveals novel aerobic deammonification in thermophilic aerobic digestion. Water Sci Technol 2003; 47(10): 131-8.
[http://dx.doi.org/10.2166/wst.2003.0557] [PMID: 12862227]

[46]   Bokare AD, Choi W. Review of iron-free Fenton-like systems for activating H_2O_2 in advanced oxidation processes. J Hazard Mater 2014; 275: 121-35.
[http://dx.doi.org/10.1016/j.jhazmat.2014.04.054] [PMID: 24857896]

[47]   Bell LC, Richardson DJ, Ferguson SJ. Periplasmic and membrane-bound respiratory nitrate reductases in *Thiosphaera pantotropha*. The periplasmic enzyme catalyzes the first step in aerobic denitrification. FEBS Lett 1990; 265(1-2): 85-7.
[http://dx.doi.org/10.1016/0014-5793(90)80889-Q] [PMID: 2365057]

[48]   Chen P, Li J, Li QX, *et al.* Simultaneous heterotrophic nitrification and aerobic denitrification by *bacterium Rhodococcus* sp. CPZ24. Bioresour Technol 2012; 116: 266-70.
[http://dx.doi.org/10.1016/j.biortech.2012.02.050] [PMID: 22531166]

[49]   Zhang Q-L, Liu Y, Ai G-M, Miao LL, Zheng HY, Liu ZP. The characteristics of a novel heterotrophic nitrification-aerobic denitrification bacterium, *Bacillus methylotrophicus* strain L7. Bioresour Technol 2012; 108 (Suppl. C): 35-44.
[http://dx.doi.org/10.1016/j.biortech.2011.12.139] [PMID: 22269053]

[50]   Ren Y-X, Yang L, Liang X. The characteristics of a novel heterotrophic nitrifying and aerobic denitrifying bacterium, *Acinetobacter junii* YB. Bioresour Technol 2014; 171: 1-9.
[http://dx.doi.org/10.1016/j.biortech.2014.08.058] [PMID: 25171329]

[51]   Kim M, Jeong S-Y, Yoon SJ, *et al.* Aerobic denitrification of *Pseudomonas putida* AD-21 at different C/N ratios. J Biosci Bioeng 2008; 106(5): 498-502.
[http://dx.doi.org/10.1263/jbb.106.498] [PMID: 19111647]

[52]   Chen Q, Ni J. Ammonium removal by Agrobacterium sp. LAD9 capable of heterotrophic nitrification-aerobic denitrification. J Biosci Bioeng 2012; 113(5): 619-23.
[http://dx.doi.org/10.1016/j.jbiosc.2011.12.012] [PMID: 22296870]

[53]   Robertson LA, Kuenen JG. Combined heterotrophic nitrification and aerobic denitrification in Thiosphaera pantotropha and other bacteria. Antonie van Leeuwenhoek 1990; 57(3): 139-52.
[http://dx.doi.org/10.1007/BF00403948] [PMID: 2181927]

[54]   Robertson LA, Kuenen JG. Aerobic denitrification--old wine in new bottles? Antonie van Leeuwenhoek 1984; 50(5-6): 525-44.
[http://dx.doi.org/10.1007/BF02386224] [PMID: 6397132]

[55]   Wrage N, Velthof G, Van Beusichem M, Oenema O. Role of nitrifier denitrification in the production of nitrous oxide. Soil Biol Biochem 2001; 33(12-13): 1723-32.
[http://dx.doi.org/10.1016/S0038-0717(01)00096-7]

CHAPTER 4

Bio Electrochemical Systems: A New Approach Towards Environmental Pollution Remediation Biotechnology

Khadijeh Jafari[1], Edris Hossienzadeh[2], Mohammad Amin Mirnasab[3,*], Ghorban Asgari[4], Ayub Ebadi Fathabad[5] and Sakine Shekoohiyan[1]

[1] *Department of Environmental Health Engineering, Faculty of Medical Sciences, Tarbiat Modares University, Tehran, Iran*

[2] *Department of Environmental Health Engineering, Social Determinants of Health Research Center, Saveh University of Medical Sciences, Saveh, Iran*

[3] *Department of Environmental Health Engineering, School of Health, Student Research Committee, Shiraz University of Medical Sciences, Shiraz, Iran*

[4] *Department of Environmental Health Engineering, Hamadan University of Medical Sciences, Hamadan, Iran*

[5] *Department of Food Hygiene and Quality Control, Faculty of Veterinary Medicine, Urmia University, Urmia, Iran*

Abstract: The entrance of various pollutants into the environment and the resulting problems led to the use of new methods for the purification of these pollutants and the recovery of energy contained therein. Nowadays, bioelectrochemical systems (BES) are considered as a new technology in the generation of electricity and wastewater treatment. Also, the use of fossil fuels, especially oil and gas, has accelerated in recent years resulting in a global energy crisis. Bio-Renewable energy is considered as one of the ways to reduce the current global warming climate. Modern production of electricity from renewable sources without much carbon dioxide emissions is much more favorable. The purpose of this chapter is to present a brief introduction on the use of bioelectrochemical systems and its various types including Microbial Fuel Cells (MFC), Enzymatic Biofuel Cells (EFC) and Microbial Electrolysis Cell (MEC).

Keywords: Bioelecterochemical Systems (BES), Bioenergy, Enzymatic Biofuel Cells, Microbial Fuel Cells, Microbial Electrolysis Cell, Wastewater Treatment.

---
* **Corresponding author Mohammad Amin Mirnasab:** Department of Environmental Health Engineering, School of Health, Student research committee, Shiraz University of Medical Sciences, Shiraz, Iran; Tel/Fax: +989140703556; E-mail: mirnasab@iran.ir

Edris Hoseinzadeh (Ed.)
All rights reserved-© 2019 Bentham Science Publishers

INTRODUCTION

The use of fossil fuels, especially oil and gas has accelerated in recent years resulting in a global energy crisis. Bio renewable energy is considered as one of the ways to reduce the current global warming climate. Most efforts were performed for the development of alternative methods for electricity production. Modern production of electricity from renewable sources without much carbon dioxide emissions is much more favorable [1, 2]. Also, the entrance of various pollutants into the environment and the resulting problems led to the use of new methods for the purification of these pollutants and the recovery of energy contained therein. Nowadays, bioelectrochemical systems (BES) are considered as a new technology in the generation of electricity and wastewater treatment [3]. In these systems, microorganisms are used to perform oxidation or reduction reactions [4] as well as direct electrical current to increase the efficiency of biological degradation [5, 6]. Therefore, BES is also known as a clean technology for wastewater treatment [7]. The purpose of this chapter is to present a brief introduction on the use of BES and their various types.

Types of Bioelectrochemical Systems

Bioelectrochemical systems (BES) are one of the most promising advanced technologies used for multifunctional applications such as Bio-energy production, wastewater treatment and synthesis of valuable chemical additives [8]. BES is a large group of biological catalytic electrochemical systems with various functions that are used to maintain and protect energy and remove substrates [9]. As said, BES is an emerging treatment technology based on microbial interaction with solid/donor electron receivers. While energy recovery from organic compounds is central to BES research [10], nutrient removal and recycling have also attracted particular attention [11]. Nitrogen is also removed by bio-electrochemical denitrification or recovered through ammonium migration through electricity production [12 - 18]. Based on the type of biocatalyst, bioelectrochemical systems are classified into two categories of microbial fuel cells (MFCs) and enzymatic fuel cells (EFCs) [16, 17]. According to the application, bioelectrochemical systems are used for microbial fuel cells (MFCs), microbial electrolytic cells (MECs), microbial solar cells (MSCs) and microbial desalination cells (MDCs) [9]. The idea of using BES as a good method for simultaneous desalination and recycling of hydrogen/energy in MDC formulation has recently been introduced [18] and later it was described by other researchers [19, 20]. Recently stripped MDCs, have been described as desalting and concentrated water chambers, which are characterized by anionic and cationic exchange membranes [21]. In fact, the term of MXC has been made for bioelectrochemical systems, which states that "X" represents different types and applications [22, 23]. Apart from the role of bioelectrochemical systems in generating electricity in MFCs and hydrogen production in MECs, BESs are used in various forms to treat some of the toughest contaminants, including industrial wastewater (including juice factories, paper, municipal wastewater, food and animal waste (for example, dyeing and removing paints, organochlorine removal, lye filtration, removal of sulfide, removal of nitrites, and removal of nitrate from wastewater) [24 - 31].

Microbial Fuel Cells

A microbial fuel cell (MFC) is a biological reactor that converts the chemical energy present in chemical bonds in organic compounds through the catalytic reactions of microorganisms under anaerobic conditions to electrical energy. For many years, this knowledge has been created that it is possible to generate electricity directly using bacteria to break organic substrates [32]. The recent energy crisis has given fresh impetus to the interests of microbial fuel cells in academic research as a way to produce electrical energy or hydrogen from biomass and without carbon emissions in the ecosystem. Microbial fuel cells can also be used in wastewater treatment plants to break down organic matter. They are also studied for applications such as biosensors, for example, sensors for monitoring of biochemical oxygen demand (BOD). The output power and coulombic efficiency (CE) were remarkably influenced by a variety of microbes in the anodic chamber of microbial fuel cell, the shape and arrangement of microbial fuel cells and functional conditions. Efforts have been made to improve the efficiency and reduce the cost of building and operating microbial fuel cell [33]. In microbial fuel cells, bacteria can be used to generate electricity, while biodegradation of organic matter and waste can be used [34, 35]. One of the biggest advantages of MFCs is that these systems can also be used at low loading rates [36]. Table **1** shows the types of microorganisms used in microbial fuel cells.

Microorganisms Function in Microbial Fuel Cells

The microbes in the anodic section of a MFC oxidize the substrates and produce electrons and protons. Carbon dioxide is produced as an oxidation product. However, there will not be any major carbon dioxide production, because carbon dioxide in renewable biomass is inherent in the process of photosynthesis from the atmosphere. In contrast to the direct combustion process, the electrons are adsorbed by anode and transferred to the cathode through the external circuit. After passing through a proton exchange membrane (PEM) or a salt bridge, protons enter the cathode section, where they are combined with oxygen in the water. The microbes in the anode component release the electrons and protons in an organic substrate oxidation sub-process. The production of electrical current is possible by separating the microbes from oxygen or any terminal receptor terminal, other than the anode, which requires an anaerobic anode section [37]. The overall reaction is substrate decomposition into carbon dioxide and water along with electricity production as a byproduct. Based on the two electrode reactions mentioned above, a biofilter reactor of a MFC can generate electricity through an electron current from the anode to the cathode in the external circuit [33]. In recent years, rapid advances have been made in the investigation of MFCs. Logan *et al.* [38] reviewed MFC designs, specifications and performance. Microbial metabolism was investigated in MFCs by Ribae and Verstriet [37]. Lovely basically examines promising microbial cell systems that are known as Benthic Unattended Generators (BUGs) that are used to provide sensory or remote sensing equipment based on microbial physiology, that in this study, the development of a microbial fuel cell that produces electricity from organic materials stored in marine sediments. Represents promising methods for generating electricity from distant environments. Further study of these systems has led to the discovery of microorganisms that store the energy needed to grow themselves, through the full oxidation of organic matter to carbon dioxide, along with the direct transfer of electrons to electrodes. It was also argued that the use of a variety of microorganisms with specific physiology in sensors and various other waste products is also likely to generate electricity, and requires significant corrective measures to generate electricity at high levels [2]. Pham *et al.* briefly outlined the advantages and disadvantages of MFCs in comparison with the conventional anaerobic digestion technology for the production of biogas as a renewable energy [30]. In another study, both the electrochemical properties of the active bacteria used in the immediate MFC and the speed limiting steps in electron transport were discussed [39].

APPLICATIONS

Electricity Generation

MFCs are capable to convert chemical energy stored in chemical compositions in biomass with the help of microorganisms. Since the chemical energy from oxidation of fuel molecules instead of heat, directly converts to electricity, the Carnot cycle with limited thermal efficiency is avoided and theoretically, a much higher conversion efficiency can be obtained (< 70%), just like traditional chemical fuel cells. Chaudhuri and Lovely [40] reported that *R. ferrireducens* could generate electricity up to 80% with electron yield. Most electrons were electrically accounted for about 89% [41]. A coulombic efficiency over than 97%, was reported during Formate oxidation with black platinum catalyst [42]. However, the production of MFC power is still very slow, which means that the rate of electron isolation is very low [43]. A feasible way to solve this problem is to store electricity in rechargeable equipment and then distribute electricity to final consumers [44]. Real robots that are energy-independent can probably be equipped with microbial fuel cells that use different fuels such as sugar, fruit, dead insects, grass and weed. The EcoBot-II robot by MFC provides its energy alone for some activities including moving, sensing, calculating and communicating [45, 46]. Local biomass production can be used to produce renewable energy for local use. The use of MFCs in space ships is also possible, because they can supply electricity by decomposing waste generated in the spaceship. Some scientists are dreaming that in the future a miniature MFC can be placed in the human body to provide energy-for implantable medical device that is supplied by the human body feeding [47].

Bio - Hydrogen

MFCs can be easily modified to generate hydrogen instead of electricity. Under normal operating conditions, the released protons are transferred to the cathode by anodic reaction to produce water combine with oxygen. Hydrogen production from protons and electrons produced by microbial metabolism in a MFC is thermodynamically undesirable [48]. MFCs can potentially produce about 8 to 9 moles of hydrogen per mole of glucose compared to 4 moles of hydrogen per mole of glucose obtained through the conventional fermentation process [48]. Biohydrogen production using MFCs does not require oxygen in the cathode chamber. As a result, the efficiency of the MFC will be improved, because the oxygen leakage is no longer relevant to the anodic chamber. Another advantage is that hydrogen can be collected and used for future use to overcome the low power characteristics of MFCs. As a result, MFCs have created a renewable source of hydrogen that can meet the overall need for hydrogen in the hydrogen economy [49].

Wastewater Treatment

Initially in 1991, it was intended that MFCs be used to wastewater treatment [50]. Urban wastewater contains several organic compounds that can be considered as fuel for MFCs. The amount of power generated by MFCs in the wastewater treatment process can potentially halve the electricity needed in a conventional treatment process that consumes a lot of electric power aerating activated sludge. MFCs remove solids between 50 to 90 percent less. Also, organic molecules such as acetate, propionate and butyrate can be completely decomposed into CO_2 and H_2O [21 - 53]. In particular, a combination of electrophiles and anodophiles is suitable for wastewater treatment because most of the organic matter can be biodegradable by a large number of organic materials. MFCs that use certain microbes have the ability to remove sulfides, as required in wastewater treatment [30]. MFCs can increase the bioelectrochemical activity of microbes during wastewater treatment, thus they have good functional stability. Due to increase scale concerns, MFCs with continuous flow and single-part and non-membrane are interested in treating wastewater [53 - 55]. Health care wastes, wastewater, industrial wastewater with high organic loading, pig wastewater and corn stalk, agricultural products and rice plants and *etc.* All of them are excellent sources of biomass for MFCs because they are rich in organic matter [8, 56, 57]. In some cases, up to 80% of COD can be eliminated [58] and also up to 80% of coulombic efficiency has been reported in some cases [59].

Biosensors

Apart from the applications mentioned above, another potential application of MFC technology is their use as a sensor for contamination analysis and monitoring and process control *in situ* [47]. The relative correlation between the efficiency of MFC and the wastewater resistance of the coulombic can allow for MFCs to be used as a BOD sensor [33].

Microbes Used in Microbial Fuel Cells

Many microorganisms have the ability to transfer obtained electrons from the metabolism of organic matter to anode. Their list is shown in Table **1** with their substrates. Marine sediments, soil, wastewater, fresh water and activated sludge, all of them are rich sources of these microorganisms [60, 61]. A number of recent publications have discussed the monitoring and characterization of microbes and the construction of a set of chromosomes for microorganisms that are capable of producing electricity from organic matter decomposition [47, 62].

Table 1. Microbes used their substrates in MFC.

Microbe	Substrate	References
Actinobacillus succinogenes	Glucose	[3]
Aeromonas hydrophila	Acetate	[4, 5]
Pseudomonas sp	Acetate	[6]
Clostridium beijerinckii	Starch, glucose, lactate, molasses	[7]
Clostridium butyricum	Starch, glucose, lactate, molasses	[4, 5]
Escherichia coli	Glucose sucrose	[8, 9]
Geobacter metallireducens	Acetate	[10]
Geobacter sulfurreducens	Acetate	[11 - 14]
Geobacteraceae, Geopsychrobacter electrodiphilus gen. nov., sp. Nov	Acetate, lactate, propionate, and butyrate	[15]
Shewenella species	Oxidation of lactate to acetate	[16]

(Table 1) cont.....

Microbe	Substrate	References
Klebsiella pneumoniae	Glucose	[17]
Lactobacillus plantarum	Glucose	[18]
Pseudomonas aeruginosa	Glucose	[19]
Rhodoferax ferrireducens	Glucose, xylose sucrose, maltose	[20]
Streptococcus lactis	Glucose	[18]
Geothrix fermentans	Lactate	[21]
Shewanella putrefaciens	Lactate, pyruvate, acetate, glucose	[22]
Alcaligenes faecalis, Enterococcus gallinarum, Pseudomonas aeruginosa	Glucose	[19]

Design of Microbial Fuel Cells

A typical MFC is composed of an anodic chamber and a cathode compartment separated by a proton exchange membrane (PEM) [82]. MFC systems are designed as single-compartment MFCs, two compartments MFCs, stacked MFCs and high-flow MFCs, each with its own advantages and disadvantages and depending on the type of process that is required, the designs are also different [33].

Performances of Microbial Fuel Cells

Effects of Operating Conditions

So far, the efficiency of laboratory MFCs has still been less than ideal. There may be several possible reasons for that. The production of power associated with a MFC is affected by many factors, including the type of microbes, type of biomass fuel and its concentration, ionic resistance, pH, temperature, reactor composition, anode and cathode capacitance function and proton exchange systems (PES). For a MFC system, the specified performance parameters should be adjusted to increase the polarity for the efficiency of a MFC increase [33].

Removal of Contaminants in Microbial Fuel Cells on the Anode

Anode is the electrode which the electron donar are undergoing oxidation reactions. Since most of the organic pollutants in the wastewater system are in reduced form, the COD index is used to measure the oxidation power of the pollutants. Therefore, the anodic chamber in MFCs has improved the oxidation power for convert COD to carbon dioxide and water [83]. Recent research on the use of anodic filtration processes to remove stagnant pollutants against decomposition such as azo dyes [84], polycyclic aromatic hydrocarbons (PAHs), inorganic wastewater containing sulfur and benzene derivatives, solid waste treatment, sediment, soil, and groundwater contaminated with other pollutants [85, 86].

On the Cathode

Cathode is the electrode where electron receptors are undergo reductive reactions. If the electrical potential of the cathodic electrode is oxidized over the threshold of the material, the process is reduced [87]. Most use of the MFC cathodic chambers for waste/wastewater treatment recently has been used to remove heavy metals such as chromium, vanadium, minerals such as ammonia, organic materials such as 3-chlorethylene, chlorobenzene, *etc.* [88 - 90].

In Integrated Systems

System integration for MFC has been widely discussed in various scientific sources as a pre-treatment step, a unit after purification and as a parallel system with the treatment system used for pollutants. Subsequently, many MFC-based hybrid systems have emerged, including sediment MFCs (SFCs), wetland MFCs (CW-MFCs), membrane bioreactor MFCs (MBR-MFCs), desalination MFCs (DS-MFCs), Electron-Fenton MFC based process, MFC combined with algae bioreactors and other modifications [86, 91 - 93].

Challenges and Solutions Presented

The use of MFCs has advantages and disadvantages in its applications. Obstacles encountered in using MFCs include high installation cost, the need for expensive metal catalysts on the cathode, production of low energy production and energy production constraints on a large scale [94]. Also, the low rate of energy production by microbes is subject to the limitations of this method. Therefore, this technology he MFC technology has to compete with methanogenic anaerobic digestion because they can utilize the same biomass in many cases for energy productions [33]. MFCs are capable of converting biomass to temperatures below 20 °C and with low concentrations of substrates, two of which are problematic for methanogenic digestion [95]. The disadvantages of the MFC method are its dependence on biofilms for the electron-free transfer [33]. An attempt was made to reduce the cost of MFCs technology, including fiber refinement, the use of air-cathode to remove the PEMs and the use of bio-coated to replace with Pt-catalytic cathodes [96 - 98]. Further research has been done in recent years on the use of MFCs for disinfection, medical applications, and generally for solving global environmental health problems, but still requires further review of this technique as high-tech technology to provide energy and environmental protection [47].

BIOFUEL CELLS

Biofuel cells and conventional fuel cells are same in terms of structure and function (Fig. **2**), while their catalysts are biological objects including enzymes, microbes, and organelles [23]. Biofuel cells made by using biological catalysts have several unique features including working at ambient temperature, catalyst-fuel specificity, usually no need to isolate the anode and cathode using a membrane, and lastly, there are large numbers of probable fuels due to the multitude of biological catalysts that can be utilized. Biofuel cell fuels (normally referred to as substrates) contain alcohols, sugars, wastewater, and biological fluids including blood, sweat, and tears. The plenty and sustainable nature of biofuel cells causes them a renewable energy choice, though it has recently been revealed that biofuel cells can even act with JP-8 aircraft fuel [24]. Table **2** shows some of fuels and enzymes used in enzymatic biofuel cells.

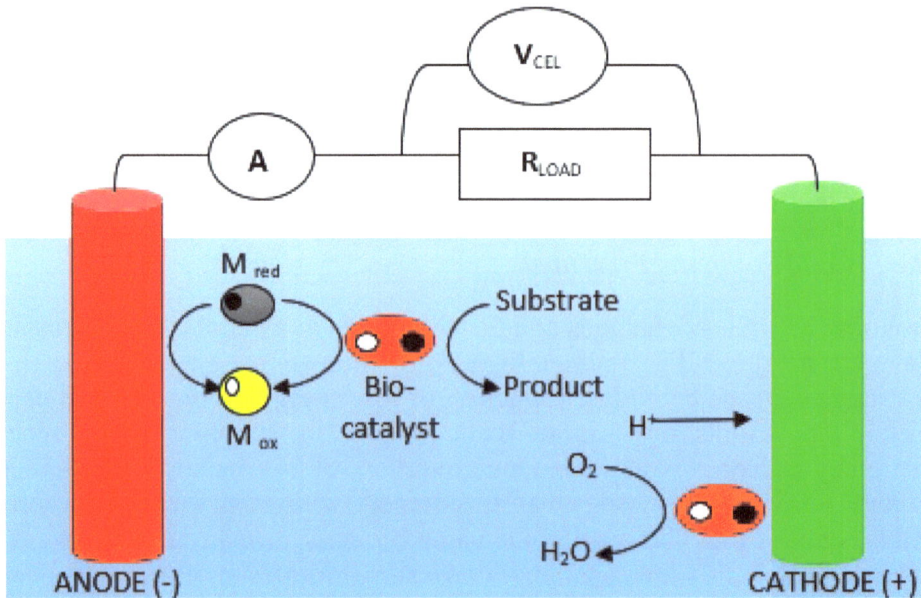

Fig. (1). Schematic of biofuel cell showing its major components and basic operation. In this example, the anodic electron transfer occurs through a mediator while at the cathode it occurs directly through a biological catalyst.

Biofuel cells are classified according to the three earlier mentioned catalyst forms including enzymatic, microbial, and organelle. Enzymatic biofuel cells include enzyme catalysts separated from a biological media and put on the anode or cathode, or both of them. Since the enzymes have been separated from their parent cells, they have ability for direct communication with mediators or electrodes, so they normally have greater power density comparing to microbial or organelle biofuel cells. Though, eliminating the enzymes from their original media also drops their sustainability [25]. On the other hand, in microbial biofuel cells, microbial catalysts are directly grown on the surface of electrodes and keep on intact over process. This significantly increases the catalysts life [26, 27], however shields enzymatic active sites, causing more difficult transferring of electron to the electrodes. Considering the third kind of biofuel cell using organelles, they are in their early step comparing with their enzymatic and microbial counterparts. Consequently, they have neither the power output of enzymatic biofuel cells nor the sustainability of microbial biofuel cells. Organelles including mitochondria, can be separated from living cells and immobilized directly on an electrode [28]. Mitochondria comprise a group of membrane-bound enzymes that procedure an electron transport chain and can directly communicate with an electrode. These membrane-bound enzymes are theoretically more steady comparing with those are in an enzymatic biofuel cell, and the mitochondria should be able to transfer electron in shorter time comparing with a MFC as cellular walls do not surround them [28]. There is a wealth of study for every of these biofuel cell groups; however the emphasis from this point forward will emphasis on enzymatic biofuel cells.

Enzymatic biofuel cells generate electricity *via* redox reactions catalyzed by enzymes. There are several stages that must happen to originate electrons in a substrate molecule for reaching the electrode, the procedure will be shown through an example of the most usual kind of enzymatic biofuel cell: a mediated glucose biofuel cell, and the main stages are considered A through E in Fig. (**2**). The procedure begins if a glucose substrate molecule in the figure, diffuses and convects from the bulk solution into the carbon nanotube (CNT)-filled polymer film whose surface is placed between two points of A and B. Diffusing over the bulk solution (point A) is directly associated with temperature (Stokes-Einstein relation) and is inversely associated with molecular weight of the substrate [29]. Furthermore, diffusion depends on the bulk solution phase (usually liquid or gas), where diffusion over a gaseous phase is normally orders of more magnitude comparing with over a liquid. Once inside the polymer film, porosity and free volume of the polymer affect diffusion of the substrate molecule, so that diffusion occurs faster because of a larger free volume [30]. The use of hydrogels is usual for immobilizing enzymes, mediators, and CNTs because of their potential to swell when hydrated. This enhances their free volume. The film immobilized shown in Fig. (**2**) cover the whole region, from the black vertical line situated between two points of A and B to the left-hand side of the figure. Though this whole region is filled with polymer, it is just signified in one place as a red polymer with ferrocene compounds connected to it. If the substrate molecule and the enzyme are nearly close to each other, they temporarily create a complex, permitting electron transfer to the enzyme reaction place (C) [31]. The enzyme must comprise, either as a part of its original structure or as an interim enzyme-cofactor complex, a redox moiety where the transformation of electron takes place. As observed in Fig. (**2**), the glucose oxidase comprises a permanently bound flavin adenine dinucleotide (FAD) moiety cofactor inside the enzyme structure that its reduction occurs by oxidation of the glucose, and then the separation of the enzyme-substrate complex occurs. According to Michaelis-Menten kinetics, the enzyme-substrate binding/release rates, product production speed, and level of both substrate and enzyme impact on substrate-enzyme electron transfer [32]. The redox moiety site inside the enzyme—whether it is close the surface or strongly embedded—also impacts electron transfer [33] *via* altering the electron tunneling distance, which has a great influence on electron transfer degree [34]. In the following stage, denoted at point D, electron of the enzyme redox place (*e.g.,* FAD) is released when the enzyme is adequately near to a moderator molecule, which must have a more redox ability comparing to the redox moiety for transferring electron. The mediating molecule shown in Fig. (**2**) is the ferrocene compound linked to the polymer. In addition to the mediator/enzyme redox potential, other main factors influencing enzyme-mediator electron transfer are mediator amount and the level of movement allowed with the polymer cross- linking procedure [110]. Though the polymer entraps enzymes/cofactors and mediators, several movements must be allowed reacting species to interact for occurring effective electron transfer. Free space between redox compounds and the polymer backbone and level of cross-linking are the main parameters for mediator movement and enzyme-mediator electron transfer [35 - 37]. At point E, transformation of electrons from the mediator to the electrode carried out directly or *via* a conductive fiber (*e.g.,* carbon nanotube, CNT) network. The mediator amount and movement are essential for transferring electron to the electrode [38], and in the case of using a conducting fiber network, the fiber density is also essential as it specifies the electrochemically active surface area of the electrode. Once transferred to the electrode, electrons transfer to the cathode *via* an exterior wire to be applied in electro reduction, which is an electron transfer procedure as same as at the anode, while it happens in the opposite direction.

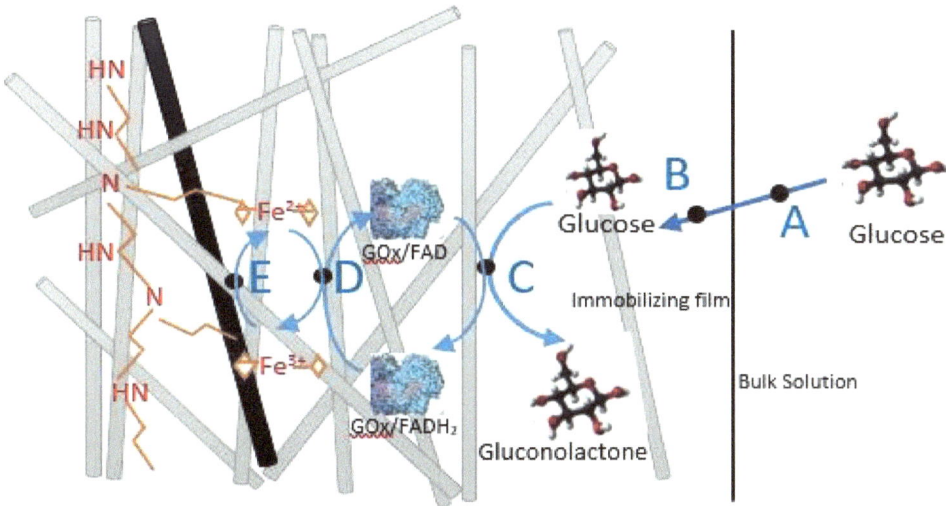

Fig. (2). The process of electron transfer from a substrate molecule to the electrode is illustrated in this schematic. Here, a glucose molecule is oxidized by the enzyme glucose oxidase (GOx), electron transfer from the enzyme to the CNT matrix is mediated by a ferrocene redox polymer.

Enzymatic Biofuel Cell Applications

One of the common use of enzymatic biofuel cells is implantable power because of their process at physiological temperature and pH and using biological fluids as fuels. Mano *et al.* reported the first living system implantation of biofuel cell in a grape [39], but the first implantation of biofuel cell in animal was reported by Cinquin *et al.* in [40]. Other examples have then followed containing those implanted in a cockroach [41, 42], snail [43], clam [44], and lobster [45]. Moreover, scientists have produced semi-implantable, or wearable, enzymatic biofuel cells like the skin patch biofuel cells presented by Wang *et al.* [46], Ogawa *et al.* [47] and the contact lens biofuel cell introduced by Reid *et al.* [48]. Though these wearable and implantable power sources do not presently create the power or have the sustainability needed for most real devices, they may eventually reach that target. Meanwhile, biofuel cells have revealed that they generate adequate power for a small biological sensor. Since daily application of wearable electronics is increased, the request for sensors running off of biological fluids will enhance. A biofuel cell can act as a power source and a sensor at a same time due to the current generated with it is directly associated with the amount of the substrates. Katz *et al.* demonstrated this concept for the first time [49], and other ones have followed [50, 51]. Furthermore, Katz *et al.* presented enzyme logic systems which pH near the electrode influences switching on or off of the bioelectro catalysis [52, 53]. This technique could sometimes be utilized for automatically supplying power of implantable devices or controlling the release of medicines in responding to chemical signals of the body. Enzymatic biofuel cells are a fascinating power source for portable electronics, since biofuels have a greater energy density comparing to rechargeable batteries [54 - 56], which causes it to be considered as a more important device volumes shrink, while power demands increase. However, biofuel cells are not recommended as feasible substitutes for lithium-ion batteries so far, although several efforts have been performed toward that goal [56, 57]. Rather, studies like microfluidic biofuel cells have made design, that is more appropriate to stationary power. Microfluidic biofuel cells, could be a cheap emergency power source that is able to use usual household objects, including table sugar (sucrose) [58] and fruit juice [59] as fuel. The microfluidic biofuel cells are appropriate for producing stationary energy because of their convective, which transport enhances mass transport to/from electrodes that can be essential for large electrodes which generate a thick layer of reaction product. Lab-on-a-chip (LOC) devices are one field of portable power that enzymatic biofuel cells may obtain a good position. These small self-contained sensors and sample processing chips can need a cheap, disposable power source, like the recent produced biofuel cells based on paper [60, 61]. The super capacitor/biofuel cell combination is one of the latest improvement in the field of enzymatic biofuel cells that tolerates mentioning [62]. The device was manufactured with super capacitive electrode materials with ability of charging and discharging using charge being provided *via* enzymatic electro catalysis.

Table 2. Fuels and enzymes used in enzymatic biofuel cells.

Fuel	Enzyme	Co-factor	Half-Cell Reaction	Natural Acceptor
Glucose	glucose oxidase, EC 1.1.3.4	FAD	glucose → glucono-1,5-lactone + $2H^+$ + $2e^-$	O_2
	glucose dehydrogenase, EC 1.1.1.47	NAD	see above	NAD
	glucose dehydrogenase, EC 1.1.5.2	PQQ	see above	quinone
	Cellobiose dehydrogenase, EC 1.1.99.18	FAD, heme	see above	acceptor
Fructose	fructose dehydrogenase, EC 1.1.99.11	FAD, heme	fructose → 5-dehydrofructose + $2H^+$ + $2e^-$	acceptor

(Table 2) cont.....

Fuel	Enzyme	Co-factor	Half-Cell Reaction	Natural Acceptor
Cellobiose	Cellobiose dehydrogenase, EC, 1.1.99.18	FAD, heme	cellobiose $\rightarrow$ cellobiono-1,5-lactone + $2H^+$ + $2e^-$	acceptor
Lactose	cellobiose dehydrogenase, EC 1.1.99.18	FAD, heme	lactose $\rightarrow$ 4-O-(galactopyranosyl)-glucono-1,5-lactone + $2H^+$ + $2e^-$	acceptor
Methanol	alcohol dehydrogenase*, EC 1.1.1.1	NAD	alcohol $\rightarrow$ aldehyde + $2H^+$ + $2e^-$	NAD
	aldehyde dehydrogenase*, EC 1.2.1.5	NAD	aldehyde + H2O $\rightarrow$ acid + $2H^+$ + $2e^-$	NAD
	formate dehydrogenase*, EC 1.2.1.2	NAD	formate $\rightarrow$ CO_2 + $2H^+$ + $2e^-$	NAD
	alcohol dehydrogenase, EC 1.1.99.8	PQQ, heme	alcohol $\rightarrow$ aldehyde + $2H^+$ + $2e^-$	acceptor
Ethanol	alcohol dehydrogenase*, EC 1.1.1.1	NAD	see above	see above
	aldehyde dehydrogenase*, EC 1.2.1.5	NAD	see above	see above
	alcohol dehydrogenase, EC 1.1.99.8	PQQ, heme	see above	see above
Glycerol	alcohol dehydrogenase*, -	PQQ, heme	alcohol $\rightarrow$ aldehyde + $2H^+$ + $2e^-$	-
	aldehyde dehydrogenase*, -	PQQ, heme	aldehyde + H_2O $\rightarrow$ acid + + $2H^+$ + $2e^-$	-
	oxalate oxidase*, EC 1.2.3.4	FAD, Mn	oxalate $\rightarrow$ $2CO_2$ + $2H^+$ + $2e^-$	O_2
Pyruvate	pyruvate dehydrogenase*, EC 1.2.4.1	NAD	pyruvate + CoA$\rightarrow$ acetylCoA + $2H^+$ + $2e^-$	NAD
Hydrogen	membrane-bound hydrogenase, -	-	$H_2 \rightarrow 2H^+$ + $2e^-$	-
*enzymes taking part in complete fuel oxidation				

Enzymatic Biofuel Cell Challenges

Most common enzymatic biofuel cells-related challenges are sustainability, electrical efficiency, performance, and energy density [23]. The first two challenges including sustainability and electrical efficiency are possibly mentioned as the most frequent challenges, maybe due to their easiest translation into system-level device needs. This section presents main parameters influencing these two challenge areas along with several remarkable efforts made for addressing them. Sustainability will be argued first, followed by electrical efficiency.

Sustainability

Implantable instruments may be the most rigorously requesting the use of biofuel cell because of their sustainability, where, for competing with available rechargeable batteries, biofuel cells must work several months, even years. As an example, pacemaker batteries presently work for more than 10 years [63] and Implantable Cardioverter-Defibrillators (ICD) for nearly 5 years [64]. Contrast this to the present state-of-the-art enzymatic biofuel cell, where a team at the University of Grenoble revealed the longest-working ones so far [65, 66]. In one trial, a glucose biofuel cell kept only 22% power following intermittent testing during one year [66]. Even in a living organism, enzymes are not very steady [67], but they are replaced over biological procedures. In an enzymatic biofuel cell, where reproduction does not happen, enzymatic sustainability declines more science the enzymes are not in their natural media and also as they may diffuse from the electrode with no adequate immobilization. This drops the enzyme level instantly nearby the electrode. Diffusion of the enzyme from the electrode can be combatted over different immobilization methods, which have improved enzymatic biofuel cell life from a few days to several weeks [66, 68 - 70]. The physical adsorption/entrapment and covalent bonding are the prime immobilization techniques. Enzymes are bonded to a conductive surface (*e.g.,* gold or carbon) *via* physical adsorption using electrostatic forces and physical entrapment commonly involves implanting the enzyme in a hydrogel or sol-gel. A conductive filler, like CNTs, can be comprised in the enzyme/gel medium for improving electron transfer process between the enzyme and the electrode [71]. Immobilizing over covalent bonding is usually utilized to bond enzymes to self-assembled monolayers (SAM) [72, 73], on an expensive metal surface, however covalent immobilization can also be performed by a cross-linking agent like glutaraldehyde [74]. Enzyme immobilization has progressed to the point where mediator sustainability is frequently a problem as much as enzyme sustainability. Like enzymes, mediators are able to diffuse from the electrode; facilitating redox polymers [75] have assisted address this problem. Scientists have started to use novel enzymes, like those separated from earlier untapped biological origins or those produced *via* protein engineering. It is expected that these methods will finally cause determining more steady enzymes and enzymes with different immobilization mechanisms for increasing biofuel cell life. One of these new enzyme origins is thermophilic bacteria that they growth in the high temperature environs like hot springs, and which yield thermophilic enzymes. The life of these enzymes is longer at ambient temperature comparing to those that must be frozen or kept in the refrigerator [76]. Protein evolution directed for biofuel cells, which first time happened in 2006 [77], mimics natural evolution through selectively spreading positive characters to enzymatic activity, sustainability, and so forth. The favorable characters can be altered over multiple iterations until the immobilized enzymes are considerably different compered enzymes separated from natural sources.

Electrical Efficiency

A somewhat unclear term, electrical efficiency is a hybrid qualitative criteria of biofuel cell current density, power density, and voltage, which amounts are associated with the definition (current and voltage produce power, PP=llll). In an optimal biofuel cell, the voltage would not be affected by the quantity of current that is drawn. In fact, the voltage usually declines severely as current enhances, and consequently, the power declines. Voltage losses are resulted by various parameters, which can be classified into three groups including losses of activation, ohmic and concentration [67]. Activation losses commonly are observed at low currents, and are a criterion of the energy barrier for transferring electron from the mediator or the enzyme to the electrode. They principally show how well the biofuel cell can eliminate electrons from the fuel and transmit them to the electrode. This kind of loss can be dropped through optimization of working conditions, selecting exact catalysts and mediators, and enhancing surface area of the electrode. Ohmic losses indicate the biofuel cell parts electrical resistance containing the resistance of the electrolyte, isolating membrane, electrodes, and any links between parts. Concentration losses take place at upper currents and are resulted by slow reactant and moving the product towards or from the electrodes. Concentration losses can occur if there is only mass transport diffusion; so, agitating or pumping the fuel solution should decline concentration losses. Latest high-power enzymatic biofuel cells were manufactured using a biological anode and cathode over 5 mAcm^{-2} and 1.5 mWcm^{-2}, with an OCV between 0.6-0.95 V [23, 78]. The OCV can be false reasoning, since it does not correspond to the highest current or the highest power; however is the voltage out electrical load connected. It is a valuable measure of how near to the anode and cathode substrate standard redox potentials the biofuel is working. From a device viewpoint, it is more beneficial to determine the voltage at highest power. As an example, in one important biofuel cell, the OCV was 0.95 V while the voltage at highest power (1.3 mWcm^{-2}) was just 0.6 V [66]. The biofuel cell voltage at highest power is a more beneficial amount than the OCV as compared with state-of-the-art enzymatic biofuel cells power and voltage with the power and voltage needed for available portable and implantable instruments. Several typical instrument requirements are as follows: pacemaker—5 to 40 mW [79] at > 2.85V [80]; contact lens glucose sensor—0.003 mW at 1.2 V [81, 82]; Fitbit ZipTM wireless activity tracker—almost 0.07 mW at 3V [46]. However, enzymatic biofuel cells may generate adequate power for several of these uses, their voltage is still rather low, and though a boost converter can be applied for increasing voltage, a pre-charging voltage of almost 1.0 V may be needed even for a very small converter [83, 84]. The voltage losses because of activation, resistance, and mass transport are the main areas that can be developed upon to enhance biofuel cell current/power yield and voltage. Earlier mention was created according DET as a method for increasing OCV, which is a sign of activating losses (over potential) needed for the anodic and cathodic reactions to take place. Moreover, it was stated earlier that biofuel cells with no an isolating membrane have a smaller amount of resistance, reducing ohmic losses. Convection is one method for decreasing losses of mass transport. This is usually performed through agitating for a macro scale biofuel cell or with pumping for a microfluidic biofuel cell.

Microfluidic biofuel cells increases mass transfer from and to the electrodes by applying forced convection. Moore *et al.* was introduced the first microfluidic enzymatic biofuel cell and after that various electrode configurations and production methods have been used to develop these types biofuel cells [10]. These biofuel cells have utilized different fuels such as malate, ethanol [85] and glucose [86]. Of specific interest to this study is the instrument introduced by Rincón *et al.* [85], which though not severely a microfluidic biofuel cell, applied a flow-through bioanode. Flow-through electrodes have been demonstrated to enhance power density and fuel application comparing with planar (flow-over) electrode cells [87], due to the fuel can attach to a greater part of the electrode surface area. In a latest example, flow-through CNT/Nafion pillar electrodes were quickly decorated in a microfluidic channel by a hydrogel Micro Stencil through following enzyme immobilization [88]. Other latest remarkable improvements in the area of microfluidic enzymatic biofuel cells contain the first ethanol biofuel cell using an enzymatic anode and cathode [89], revealed application of low-cost, commercially accessible gold covered optical fibers for scalable microfluidic electrodes [90], and with an enzymatic microfluidic biofuel cell (along with a boost converter) for powering a heat sensor and wirelessly transfer data [91].

Introduction of MEC

One of the main universal issues nowadays is for satisfying the fast increasing energy demands. The current energy system has two key problems; first, the energy is generated mostly using fossil fuels, which have restricted resources and second, the application of hydrocarbon fuels results dangerous environmental contamination because of the emitting enormous amount of CO_2. That is the reason for the need to use substitute power sources, which can slowly substitute the old-style energy fuels, is generally argued [92]. Most of researchers believe that the main power source in the future is the hydrogen [93, 94], because it has three times more caloricity as compared with petrol and it has no corrosive impact on the metals. However, its widespread use is limited by its high cost generation procedure [95]. One of the two most usual approaches presently applied for producing hydrogen is conventional electrolysis. Hydrogen generation through electrolysis is not associated with emission of CO_2. The inadequacy of this technique is great consumption of electricity. A partial solution to overcome this problem is microbial electrolysis cells (MECs). A MEC is an electrolyzer, which performs oxidation of organic substance at the anode, while the cathode performs the abiotic reduction of water in the normal manner [96]. The over potential is decreased by applying the microorganisms in electrolysis systems and the electron transfer occurs easier and the amount of electricity needed for the electrolysis is decreased. Ideally, the hydrogen development on the cathode requires a potential of ECAT = − 0.41 V (*vs.* SHE). Most Microbial fuel cells (MFCs) have the anode potential of about EAN = − 0.30V (*vs.* SHE). Consequently, the least total cell voltage required is E= -0.11V [97].

$$E= E_{CAT}\text{-} E_{AN} = (\text{-}0.41) – (\text{-}0.30)= \text{-}0.11\ V \tag{1}$$

The main thing is that the MECs utilize substrates from renewable sources and have great conversion performance. That is the reason for using MEC in local power plants as a suitable method and also for treating wastewater and generating hydrogen at a same time. The development MECs show an interesting modern area in the field of environmental biotechnological study that is developing usual wastewater as a fuel source for producing hydrogen [94]. The MEC has been widely examined and improved in recent five years. In order to improve this original technique, it is important to generalize the achievements.

MEC

Structure and Operating Principle

The microbial electrolysis cell (MECs) is a technology for hydrogen generation strictly associated with MFCs. Whilst MFC's generate an electric current through the microbial disintegration of organic materials, MEC's partly reverse the procedure to produce hydrogen or methane from organic compounds by means of an electric current. The anode procedure of MEC and MFC is similar and the cathode procedure is of MEC and water electrolyzer is similar [98]. Technically, MEC can be separated into three main parts: anode chamber, cathode chamber and separator [99].

In an MEC as illustrated in Fig. (**3**), organic materials are broken down into CO_2, electrons and protons by electrochemically active microbes grown on the anode surface. The electrons and protons flow over the exterior electric circuit and the electrolyte, respectively, and merge at the cathode to produce hydrogen. A microbial biofilm on the electrode as an electrocatalyst supports the oxidation process at the anode. MEC suggests to plan particular microbial biofilms for developing MEC anodes through identification of new sources of inoculum, adapting the microbial population and determination of the structural properties of biofilm that will improve its electroactive features [100]. The most studied microbial cultures for using in MECs are Archaea, the single-celled *cyanobacterium Cyanothece 51142* [100], Dechlorinating bacteria (*Dehalococcoides* spp. and *Desulfitobacterium* spp.) and also methanogens and homoacetogens microorganisms [101]. Bio waste and wastewater cause fast profits and the maximum possibility for the effectiveness of MEC method. Abundant and renewable cellulosic biomass can possibly generate sufficient hydrogen to utilize in transporting systems and industries [97]. The identity of the particular microorganisms defines the products and the performance of the MEC. Electrogenic microorganisms produce electrons and protons through consumption of an energy source (for example acetic acid), and generate an electrical potential of up to 0.3 V. In a common MFC, this voltage is applied for producing electrical power. In a MEC, an external power source is used to provide an extra voltage to the cell. The hybrid voltage is adequate to decrease protons and generate hydrogen gas. The performance of hydrogen generation is affected by organic materials utilized. Lactic and acetic acid attain 82% performance, while the amounts for non-pretreated cellulose or glucose are near to 63%. The performance of usual water electrolysis is 60 to 70%. As MEC's change unfeasible biomass into usable hydrogen, they can generate 144% more usable energy as compared with their electrical energy consumption. Moreover, MEC's can generate methane through a corresponding mechanism depending on the organisms offered at the anode [102].

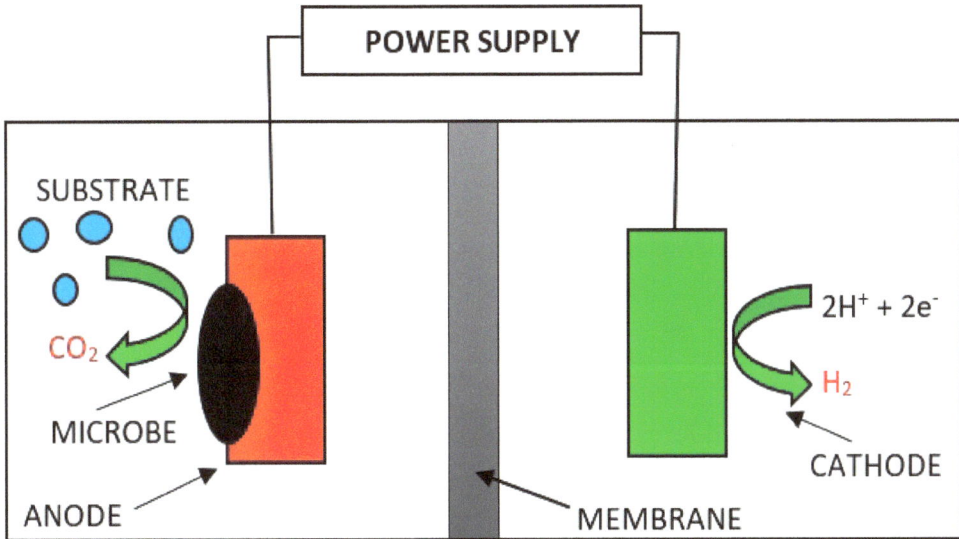

Fig. (3). The working principle of a microbial electrolysis cell.

Electrode Catalyst Materials

Similar materials can be used for anode in MEC and MFC, including carbon cloth, carbon paper, graphite felt, graphite granules or graphite brushes. The improvement of highly effective anode materials is important to enhance the current output of microbial electrochemical cells. Au and Pd NPs adorned graphite anodes were improved and assessed in a recently planned multi-anode microbial electrolysis cell (MEC). In contrast, no obvious relation was between the density of current and the density of particle on the basis of area fraction and number of particles. These outcomes revealed that Nano-decoration can strongly improve the efficiency of microbial anodes but the chemical composition, dimension and form of the nanoparticles showed the degree of improvement [103]. The application of carbon cloth or thin graphite felt as anode materials may offer great benefits compared to other materials for scale-up. The best cathode catalyst material for MEC is platinum, but it is too costly [98, 104]. The high price of platinum encourages researchers to study bio-cathodes as a substitute. Moreover, extensive research has been performed on costly metal-free catalysts for MEC. Inexpensive materials as stainless steel and carbon based NiMo-, NiW-nanocomposites demonstrated high efficiency as cathode in MEC [105]. Penn State researchers have reported a method for replacing the platinum catalyst in their hydrogen producing microbial electrolysis cells using stainless steel brushes with no loss in the performance. The trapped hydrogen also stays for longer time in the reactor and so, it is accessible for microbes to consume [106]. The disadvantages of extending these outcomes to MECs is that they have been only studied under extremely acid or alkaline conditions [106, 107]. Study of cathode materials that are effective catalysts at neutral pH is required [96].

General MEC Design

MEC can be improved as two systems, a single-chamber system or a two-chamber system. A MEC lacking a membrane can generate great hydrogen and energy efficiencies. Working with no membrane can permit for more economical and simpler plans [97]. Application of a membrane in two-chamber MEC systems can decrease the quantity of contaminations in the biogas leading to increase in the interior resistance. Removal of the membrane in single-chamber systems simplifies the reactor design, but enhances the probability for consuming hydrogen by methanogenesis bacteria [96].

Conclusion (Perspectives of Application)

Studies on the original and new MEC technology are at the initial stage. Since the reactions related to bioanode are much more widely investigated in MFC, the growth of economical cathodes for applying under almost neutral pH and low temperature conditions as well as minimizing the interior resistance through improving the cell design, is essential for practical use of the MEC and its commercialization. MECs have potential to be utilized in various fields. MEC is mainly used for producing hydrogen, which can be utilized in diverse aims. Another use of MEC is in efficient wastewater treatment. MEC can be used for converting waste organic matter into a valuable energy source [108].

CONSENT FOR PUBLICATION

Not applicable.

REFERENCES

[1]   Davis F, Higson SP. Biofuel cells--recent advances and applications. Biosens Bioelectron 2007; 22(7): 1224-35.
        [http://dx.doi.org/10.1016/j.bios.2006.04.029] [PMID: 16781864]

[2]   Lovley DR. Microbial fuel cells: novel microbial physiologies and engineering approaches. Curr Opin Biotechnol 2006; 17(3): 327-32.
        [http://dx.doi.org/10.1016/j.copbio.2006.04.006] [PMID: 16679010]

[3]   Park DH, Zeikus JG. Electricity generation in microbial fuel cells using neutral red as an electronophore. Appl Environ Microbiol 2000; 66(4): 1292-7.
        [http://dx.doi.org/10.1128/AEM.66.4.1292-1297.2000] [PMID: 10742202]

[4]   Pham CA, Jung SJ, Phung NT, *et al.* A novel electrochemically active and Fe(III)-reducing bacterium phylogenetically related to *Aeromonas hydrophila*, isolated from a microbial fuel cell. FEMS Microbiol Lett 2003; 223(1): 129-34.
        [http://dx.doi.org/10.1016/S0378-1097(03)00354-9] [PMID: 12799011]

[5]   Park HS, Kim BH, Kim HS, *et al.* A novel electrochemically active and Fe (III)-reducing bacterium phylogenetically related to Clostridium butyricum isolated from a microbial fuel cell. Anaerobe 2001; 7(6): 297-306.
[http://dx.doi.org/10.1006/anae.2001.0399]

[6]   Pham TH, Boon N, Aelterman P, *et al.* Metabolites produced by Pseudomonas sp. enable a Gram-positive bacterium to achieve extracellular electron transfer. Appl Microbiol Biotechnol 2008; 77(5): 1119-29.
[http://dx.doi.org/10.1007/s00253-007-1248-6] [PMID: 17968538]

[7]   Niessen J, Schröder U, Scholz F. Exploiting complex carbohydrates for microbial electricity generation–a bacterial fuel cell operating on starch. Electrochem Commun 2004; 6(9): 955-8.
[http://dx.doi.org/10.1016/j.elecom.2004.07.010]

[8]   Zhang T, Cui C, Chen S, *et al.* A novel mediatorless microbial fuel cell based on direct biocatalysis of *Escherichia coli.* Chem Commun (Camb) 2007; (31): 3306-06.
[PMID: 16718321]

[9]   Zhang T, Cui C, Chen S, *et al.* A novel mediatorless microbial fuel cell based on direct biocatalysis of Escherichia coli. Chem Commun (Camb) 2006; (21): 2257-9.
[http://dx.doi.org/10.1039/b600876c] [PMID: 16718321]

[10]   Moore CM, Minteer SD, Martin RS. Microchip-based ethanol/oxygen biofuel cell. Lab Chip 2005; 5(2): 218-25.
[http://dx.doi.org/10.1039/b412719f] [PMID: 15672138]

[11]   Reguera G, Nevin KP, Nicoll JS, Covalla SF, Woodard TL, Lovley DR. Biofilm and nanowire production leads to increased current in Geobacter sulfurreducens fuel cells. Appl Environ Microbiol 2006; 72(11): 7345-8.
[http://dx.doi.org/10.1128/AEM.01444-06] [PMID: 16936064]

[12]   Holmes DE, Mester T, O'Neil RA, *et al.* Genes for two multicopper proteins required for Fe(III) oxide reduction in Geobacter sulfurreducens have different expression patterns both in the subsurface and on energy-harvesting electrodes. Microbiology 2008; 154(Pt 5): 1422-35.
[http://dx.doi.org/10.1099/mic.0.2007/014365-0] [PMID: 18451051]

[13]   Izallalen M, Mahadevan R, Burgard A, *et al.* Geobacter sulfurreducens strain engineered for increased rates of respiration. Metab Eng 2008; 10(5): 267-75.
[http://dx.doi.org/10.1016/j.ymben.2008.06.005] [PMID: 18644460]

[14]   Ishii S, Watanabe K, Yabuki S, Logan BE, Sekiguchi Y. Comparison of electrode reduction activities of *Geobacter sulfurreducens* and an enriched consortium in an air-cathode microbial fuel cell. Appl Environ Microbiol 2008; 74(23): 7348-55.
[http://dx.doi.org/10.1128/AEM.01639-08] [PMID: 18836002]

[15]   Holmes DE, Nicoll JS, Bond DR, Lovley DR. Potential role of a novel psychrotolerant member of the family Geobacteraceae, Geopsychrobacter electrodiphilus gen. nov., sp. nov., in electricity production by a marine sediment fuel cell. Appl Environ Microbiol 2004; 70(10): 6023-30.
[http://dx.doi.org/10.1128/AEM.70.10.6023-6030.2004] [PMID: 15466546]

[16]   Marsili E, Baron DB, Shikhare ID, Coursolle D, Gralnick JA, Bond DR. Shewanella secretes flavins that mediate extracellular electron transfer. Proc Natl Acad Sci USA 2008; 105(10): 3968-73.
[http://dx.doi.org/10.1073/pnas.0710525105] [PMID: 18316736]

[17]   Menicucci J, Beyenal H, Marsili E, Veluchamy RA, Demir G, Lewandowski Z. Procedure for determining maximum sustainable power generated by microbial fuel cells. Environ Sci Technol 2006; 40(3): 1062-8.
[http://dx.doi.org/10.1021/es051180l] [PMID: 16509358]

[18]   Vega CA, Fernández I. Mediating effect of ferric chelate compounds in microbial fuel cells with *Lactobacillus plantarum, Streptococcus lactis*, and *Erwinia dissolvens*. Bioelectrochem Bioenerg

1987; 17(2): 217-22.
[http://dx.doi.org/10.1016/0302-4598(87)80026-0]

[19]   Rabaey K, Boon N, Siciliano SD, Verhaege M, Verstraete W. Biofuel cells select for microbial consortia that self-mediate electron transfer. Appl Environ Microbiol 2004; 70(9): 5373-82.
[http://dx.doi.org/10.1128/AEM.70.9.5373-5382.2004] [PMID: 15345423]

[20]   Liu ZD, Lian J, Du ZW, Li H-R. [Construction of sugar-based microbial fuel cells by dissimilatory metal reduction bacteria]. Sheng Wu Gong Cheng Xue Bao 2006; 22(1): 131-7.
[http://dx.doi.org/10.1016/S1872-2075(06)60010-1] [PMID: 16572853]

[21]   Nevin KP, Lovley DR. Mechanisms for accessing insoluble Fe(III) oxide during dissimilatory Fe(III) reduction by Geothrix fermentans. Appl Environ Microbiol 2002; 68(5): 2294-9.
[http://dx.doi.org/10.1128/AEM.68.5.2294-2299.2002] [PMID: 11976100]

[22]   Myers CR, Myers JM. Localization of cytochromes to the outer membrane of anaerobically grown Shewanella putrefaciens MR-1. J Bacteriol 1992; 174(11): 3429-38.
[http://dx.doi.org/10.1128/jb.174.11.3429-3438.1992] [PMID: 1592800]

[23]   Rasmussen M, Abdellaoui S, Minteer SD. Enzymatic biofuel cells: 30 years of critical advancements. Biosens Bioelectron 2016; 76: 91-102.
[http://dx.doi.org/10.1016/j.bios.2015.06.029] [PMID: 26163747]

[24]   Ulyanova Y, Arugula MA, Rasmussen M, *et al.* Bioelectrocatalytic oxidation of alkanes in a JP-8 enzymatic biofuel cell. ACS Catal 2014; 4(12): 4289-94.
[http://dx.doi.org/10.1021/cs500802d]

[25]   Moehlenbrock MJ, Minteer SD. Extended lifetime biofuel cells. Chem Soc Rev 2008; 37(6): 1188-96.
[http://dx.doi.org/10.1039/b708013c] [PMID: 18497931]

[26]   Zhang F, Ge Z, Grimaud J, Hurst J, He Z. Long-term performance of liter-scale microbial fuel cells treating primary effluent installed in a municipal wastewater treatment facility. Environ Sci Technol 2013; 47(9): 4941-8.
[http://dx.doi.org/10.1021/es400631r] [PMID: 23517192]

[27]   Aulenta F, Tocca L, Verdini R, Reale P, Majone M. Dechlorination of trichloroethene in a continuous-flow bioelectrochemical reactor: effect of cathode potential on rate, selectivity, and electron transfer mechanisms. Environ Sci Technol 2011; 45(19): 8444-51.
[http://dx.doi.org/10.1021/es202262y] [PMID: 21877695]

[28]   Arechederra R, Minteer SD. Organelle-based biofuel cells: immobilized mitochondria on carbon paper electrodes. Electrochim Acta 2008; 53(23): 6698-703.
[http://dx.doi.org/10.1016/j.electacta.2008.01.074]

[29]   Valencia DP, González FJ. Understanding the linear correlation between diffusion coefficient and molecular weight. A model to estimate diffusion coefficients in acetonitrile solutions. Electrochem Commun 2011; 13(2): 129-32.
[http://dx.doi.org/10.1016/j.elecom.2010.11.032]

[30]   George SC, Thomas S. Transport phenomena through polymeric systems. Prog Polym Sci 2001; 26(6): 985-1017.
[http://dx.doi.org/10.1016/S0079-6700(00)00036-8]

[31]   Faulkner LL, Bard AJ. Electrochemical Methods and Applications. John Wiley & Sons 2001.

[32]   Bisswanger H. Enzyme kinetics: principles and methods. John Wiley & Sons 2017.
[http://dx.doi.org/10.1002/9783527806461]

[33]   Roth JP, Klinman JP. Catalysis of electron transfer during activation of O_2 by the flavoprotein glucose oxidase. Proc Natl Acad Sci USA 2003; 100(1): 62-7.
[http://dx.doi.org/10.1073/pnas.252644599] [PMID: 12506204]

[34]   Chidsey CE. Free energy and temperature dependence of electron transfer at the metal-electrolyte

interface. Science 1991; 251(4996): 919-22.
[http://dx.doi.org/10.1126/science.251.4996.919] [PMID: 17847385]

[35]   Merchant SA, Meredith MT, Tran TO, *et al.* Effect of mediator spacing on electrochemical and enzymatic response of ferrocene redox polymers. J Phys Chem C 2010; 114(26): 11627-34.
[http://dx.doi.org/10.1021/jp911188r]

[36]   Mao F, Mano N, Heller A. Long tethers binding redox centers to polymer backbones enhance electron transport in enzyme "Wiring" hydrogels. J Am Chem Soc 2003; 125(16): 4951-7.
[http://dx.doi.org/10.1021/ja029510e] [PMID: 12696915]

[37]   Soukharev V, Mano N, Heller A. A four-electron $O_{(2)}$-electroreduction biocatalyst superior to platinum and a biofuel cell operating at 0.88 V. J Am Chem Soc 2004; 126(27): 8368-9.
[http://dx.doi.org/10.1021/ja0475510] [PMID: 15237980]

[38]   Cao L. Immobilised enzymes: science or art? Curr Opin Chem Biol 2005; 9(2): 217-26.
[http://dx.doi.org/10.1016/j.cbpa.2005.02.014] [PMID: 15811808]

[39]   Mano N, Mao F, Heller A. Characteristics of a miniature compartment-less glucose-O_2 biofuel cell and its operation in a living plant. J Am Chem Soc 2003; 125(21): 6588-94.
[http://dx.doi.org/10.1021/ja0346328] [PMID: 12785800]

[40]   Cinquin P, Gondran C, Giroud F, *et al.* A glucose biofuel cell implanted in rats. PLoS One 2010; 5(5)e10476
[http://dx.doi.org/10.1371/journal.pone.0010476] [PMID: 20454563]

[41]   Rasmussen M, Ritzmann RE, Lee I, Pollack AJ, Scherson D. An implantable biofuel cell for a live insect. J Am Chem Soc 2012; 134(3): 1458-60.
[http://dx.doi.org/10.1021/ja210794c] [PMID: 22239249]

[42]   Schwefel J, Ritzmann RE, Lee IN, *et al.* Wireless communication by an autonomous self-powered cyborg insect. J Electrochem Soc 2015; 161(13): H3113-16.
[http://dx.doi.org/10.1149/2.0171413jes]

[43]   Halámková L, Halámek J, Bocharova V, Szczupak A, Alfonta L, Katz E. Implanted biofuel cell operating in a living snail. J Am Chem Soc 2012; 134(11): 5040-3.
[http://dx.doi.org/10.1021/ja211714w] [PMID: 22401501]

[44]   Szczupak A, Halámek J, Halámková L, *et al.* Living battery–biofuel cells operating *in vivo* in clams. Energy Environ Sci 2012; 5(10): 8891-5.
[http://dx.doi.org/10.1039/c2ee21626d]

[45]   MacVittie K, Halámek J, Halámková L, *et al.* From "cyborg" lobsters to a pacemaker powered by implantable biofuel cells. Energy Environ Sci 2013; 6(1): 81-6.
[http://dx.doi.org/10.1039/C2EE23209J]

[46]   Reid RC. cdmbuntu.lib.utah.edu2016.

[47]   Ogawa Y, Kato K, Miyake T, *et al.* Organic transdermal iontophoresis patch with built-in biofuel cell. Adv Healthc Mater 2015; 4(4): 506-10.
[http://dx.doi.org/10.1002/adhm.201400457] [PMID: 25402232]

[48]   Reid RC, Minteer SD, Gale BK. Contact lens biofuel cell tested in a synthetic tear solution. Biosens Bioelectron 2015; 68: 142-8.
[http://dx.doi.org/10.1016/j.bios.2014.12.034] [PMID: 25562741]

[49]   Katz E, Bückmann AF, Willner I. Self-powered enzyme-based biosensors. J Am Chem Soc 2001; 123(43): 10752-3.
[http://dx.doi.org/10.1021/ja0167102] [PMID: 11674014]

[50]   Kakehi N, Yamazaki T, Tsugawa W, Sode K. A novel wireless glucose sensor employing direct electron transfer principle based enzyme fuel cell. Biosens Bioelectron 2007; 22(9-10): 2250-5.
[http://dx.doi.org/10.1016/j.bios.2006.11.004] [PMID: 17166711]

[51]   Valdés-Ramírez G, Li Y-C, Kim J, *et al.* Microneedle-based self-powered glucose sensor. Electrochem Commun 2014; 47: 58-62.
[http://dx.doi.org/10.1016/j.elecom.2014.07.014]

[52]   Zhou J, Tam TK, Pita M, Ornatska M, Minko S, Katz E. Bioelectrocatalytic system coupled with enzyme-based biocomputing ensembles performing boolean logic operations: approaching "smart" physiologically controlled biointerfaces. ACS Appl Mater Interfaces 2009; 1(1): 144-9.
[http://dx.doi.org/10.1021/am800088d] [PMID: 20355766]

[53]   Amir L, Tam TK, Pita M, Meijler MM, Alfonta L, Katz E. Biofuel cell controlled by enzyme logic systems. J Am Chem Soc 2009; 131(2): 826-32.
[http://dx.doi.org/10.1021/ja8076704] [PMID: 19105750]

[54]   Wu X, Ge J, Yang C, Hou M, Liu Z. Facile synthesis of multiple enzyme-containing metal-organic frameworks in a biomolecule-friendly environment. Chem Commun (Camb) 2015; 51(69): 13408-11.
[http://dx.doi.org/10.1039/C5CC05136C] [PMID: 26214658]

[55]   Abdellaoui S, Hickey DP, Stephens AR, Minteer SD. Recombinant oxalate decarboxylase: enhancement of a hybrid catalytic cascade for the complete electro-oxidation of glycerol. Chem Commun (Camb) 2015; 51(76): 14330-3.
[http://dx.doi.org/10.1039/C5CC06131H] [PMID: 26271633]

[56]   Zhu Z, Kin Tam T, Sun F, You C, Percival Zhang YH. A high-energy-density sugar biobattery based on a synthetic enzymatic pathway. Nat Commun 2014; 5: 3026.
[http://dx.doi.org/10.1038/ncomms4026] [PMID: 24445859]

[57]   Sakai H, Nakagawa T, Tokita Y, *et al.* A high-power glucose/oxygen biofuel cell operating under quiescent conditions. Energy Environ Sci 2009; 2(1): 133-8.
[http://dx.doi.org/10.1039/B809841G]

[58]   Hickey DP, Giroud F, Schmidtke DW, Glatzhofer DT, Minteer SD. Enzyme cascade for catalyzing sucrose oxidation in a biofuel cell. ACS Catal 2013; 3(12): 2729-37.
[http://dx.doi.org/10.1021/cs4003832]

[59]   Liu Y, Dong S. A biofuel cell harvesting energy from glucose-air and fruit juice-air. Biosens Bioelectron 2007; 23(4): 593-7.
[http://dx.doi.org/10.1016/j.bios.2007.06.002] [PMID: 17720474]

[60]   Villarrubia CWN, Lau C, Ciniciato GP, *et al.* Practical electricity generation from a paper based biofuel cell powered by glucose in ubiquitous liquids. Electrochem Commun 2014; 45: 44-7.
[http://dx.doi.org/10.1016/j.elecom.2014.05.010]

[61]   Zhang L, Zhou M, Wen D, Bai L, Lou B, Dong S. Small-size biofuel cell on paper. Biosens Bioelectron 2012; 35(1): 155-9.
[http://dx.doi.org/10.1016/j.bios.2012.02.035] [PMID: 22417872]

[62]   Agnes C, Holzinger M, Le Goff A, *et al.* Supercapacitor/biofuel cell hybrids based on wired enzymes on carbon nanotube matrices: autonomous reloading after high power pulses in neutral buffered glucose solutions. Energy Environ Sci 2014; 7(6): 1884-8.
[http://dx.doi.org/10.1039/C3EE43986K]

[63]   Mallela VS, Ilankumaran V, Rao NS. Trends in cardiac pacemaker batteries. Indian Pacing Electrophysiol J 2004; 4(4): 201-12.
[PMID: 16943934]

[64]   Horlbeck FW, Mellert F, Kreuz J, Nickenig G, Schwab JO. Real-world data on the lifespan of implantable cardioverter-defibrillators depending on manufacturers and the amount of ventricular pacing. J Cardiovasc Electrophysiol 2012; 23(12): 1336-42.
[http://dx.doi.org/10.1111/j.1540-8167.2012.02408.x] [PMID: 22909190]

[65]   El Ichi S, Zebda A, Alcaraz J-P, *et al.* Bioelectrodes modified with chitosan for long-term energy supply from the body. Energy Environ Sci 2015; 8(3): 1017-26.

[http://dx.doi.org/10.1039/C4EE03430A]

[66] Zebda A, Gondran C, Le Goff A, Holzinger M, Cinquin P, Cosnier S. Mediatorless high-power glucose biofuel cells based on compressed carbon nanotube-enzyme electrodes. Nat Commun 2011; 2: 370.
[http://dx.doi.org/10.1038/ncomms1365] [PMID: 21712818]

[67] Osman MH, Shah AA, Walsh FC. Recent progress and continuing challenges in bio-fuel cells. Part I: enzymatic cells. Biosens Bioelectron 2011; 26(7): 3087-102.
[http://dx.doi.org/10.1016/j.bios.2011.01.004] [PMID: 21295964]

[68] Karimi A, Othman A, Uzunoglu A, Stanciu L, Andreescu S. Graphene based enzymatic bioelectrodes and biofuel cells. Nanoscale 2015; 7(16): 6909-23.
[http://dx.doi.org/10.1039/C4NR07586B] [PMID: 25832672]

[69] Kim J, Jia H, Wang P. Challenges in biocatalysis for enzyme-based biofuel cells. Biotechnol Adv 2006; 24(3): 296-308.
[http://dx.doi.org/10.1016/j.biotechadv.2005.11.006] [PMID: 16403612]

[70] Leech D, Kavanagh P, Schuhmann W. Enzymatic fuel cells: Recent progress. Electrochim Acta 2012; 84: 223-34.
[http://dx.doi.org/10.1016/j.electacta.2012.02.087]

[71] Reid RC, Giroud F, Minteer SD, Gale BK. Enzymatic biofuel cell with a flow-through toray paper bioanode for improved fuel utilization. J Electrochem Soc 2013; 160(9): H612-19.
[http://dx.doi.org/10.1149/2.099309jes]

[72] Ammam M, Fransaer J. Gold electrode modified with a self-assembled glucose oxidase and 2, 6-pyridinedicarboxylic acid as novel glucose bioanode for biofuel cells. J Power Sources 2014; 257: 272-9.
[http://dx.doi.org/10.1016/j.jpowsour.2014.02.007]

[73] Willner I, Heleg-Shabtai V, Blonder R, *et al.* Electrical wiring of glucose oxidase by reconstitution of FAD-modified monolayers assembled onto Au-electrodes. J Am Chem Soc 1996; 118(42): 10321-2. [JACS].
[http://dx.doi.org/10.1021/ja9608611]

[74] Sheldon R. Cross-linked enzyme aggregates (CLEA® s): stable and recyclable biocatalysts.Portland Press Limited 2007.

[75] Merchant SA, Glatzhofer DT, Schmidtke DW. Effects of electrolyte and pH on the behavior of cross-linked films of ferrocene-modified poly(ethylenimine). Langmuir 2007; 23(22): 11295-302.
[http://dx.doi.org/10.1021/la701521s] [PMID: 17902716]

[76] Beneyton T, El Harrak A, Griffiths A, Hellwig P, Taly V. Immobilization of CotA, an extremophilic laccase from Bacillus subtilis, on glassy carbon electrodes for biofuel cell applications. Electrochem Commun 2011; 13(1): 24-7.
[http://dx.doi.org/10.1016/j.elecom.2010.11.003]

[77] Zhu Z, Momeu C, Zakhartsev M, Schwaneberg U. Making glucose oxidase fit for biofuel cell applications by directed protein evolution. Biosens Bioelectron 2006; 21(11): 2046-51.
[http://dx.doi.org/10.1016/j.bios.2005.11.018] [PMID: 16388946]

[78] Yamagiwa K, Ikeda Y, Yasueda K, *et al.* Improvement of electrochemical performance of bilirubin oxidase modified gas diffusion biocathode by hydrophilic binder. J Electrochem Soc 2015; 162(14): F1425-30.
[http://dx.doi.org/10.1149/2.0491514jes]

[79] Cosnier S, Le Goff A, Holzinger M. Towards glucose biofuel cells implanted in human body for powering artificial organs. Electrochem Commun 2014; 38: 19-23.
[http://dx.doi.org/10.1016/j.elecom.2013.09.021]

[80] Hogarth AJ, Artis NJ, Sivananthan UM, Pepper CB. Cardiac magnetic resonance imaging of a patient

with an magnetic resonance imaging conditional permanent pacemaker. Heart Int 2011; 6(2)e19
[http://dx.doi.org/10.4081/hi.2011.e19] [PMID: 22355486]

[81]   Yao H, Liao Y, Lingley A, *et al.* A contact lens with integrated telecommunication circuit and sensors for wireless and continuous tear glucose monitoring. J Micromech Microeng 2012; 22(7)075007
[http://dx.doi.org/10.1088/0960-1317/22/7/075007]

[82]   Liao Y-T, Yao H, Lingley A, Parviz B, Otis BP. A CMOS Glucose Sensor for Wireless Contact-Lens Tear Glucose Monitoring. IEEE J SOLID-ST CIRC 2012; 47(1): 335-44.
[http://dx.doi.org/10.1109/JSSC.2011.2170633]

[83]   Erbay C, Carreon-Bautista S, Sanchez-Sinencio E, Han A. High performance monolithic power management system with dynamic maximum power point tracking for microbial fuel cells. Environ Sci Technol 2014; 48(23): 13992-9.
[http://dx.doi.org/10.1021/es501426j] [PMID: 25365216]

[84]   Carreon-Bautista S, Erbay C, Han A, Sanchez-Sinencio E. An Inductorless DC–DC Converter for an Energy Aware Power Management Unit Aimed at Microbial Fuel Cell Arrays. IEEE J EM SEL TOP P 2015; 3(4): 1109-21.
[http://dx.doi.org/10.1109/JESTPE.2015.2398851]

[85]   Rincón RA, Lau C, Luckarift HR, *et al.* Enzymatic fuel cells: integrating flow-through anode and air-breathing cathode into a membrane-less biofuel cell design. Biosens Bioelectron 2011; 27(1): 132-6.
[http://dx.doi.org/10.1016/j.bios.2011.06.029] [PMID: 21775124]

[86]   González-Guerrero MJ, Esquivel JP, Sánchez-Molas D, *et al.* Membraneless glucose/O_2 microfluidic enzymatic biofuel cell using pyrolyzed photoresist film electrodes. Lab Chip 2013; 13(15): 2972-9.
[http://dx.doi.org/10.1039/c3lc50319d] [PMID: 23719742]

[87]   Kjeang E, Michel R, Harrington DA, Djilali N, Sinton D. A microfluidic fuel cell with flow-through porous electrodes. J Am Chem Soc 2008; 130(12): 4000-6.
[http://dx.doi.org/10.1021/ja078248c] [PMID: 18314983]

[88]   Choi SD, Choi JH, Kim YH, *et al.* Enzyme immobilization on microelectrode arrays of CNT/Nafion nanocomposites fabricated using hydrogel microstencils. Microelectron Eng 2015; 141: 193-7.
[http://dx.doi.org/10.1016/j.mee.2015.03.045]

[89]   Selloum D, Tingry S, Techer V, *et al.* Optimized electrode arrangement and activation of bioelectrodes activity by carbon nanoparticles for efficient ethanol microfluidic biofuel cells. J Power Sources 2014; 269: 834-40.
[http://dx.doi.org/10.1016/j.jpowsour.2014.07.052]

[90]   Desmaële D, Renaud L, Tingry S. Gold coated optical fibers as three-dimensional electrodes for microfluidic enzymatic biofuel cells: Toward geometrically enhanced performance. Biomicrofluidics 2015; 9(4)041102
[http://dx.doi.org/10.1063/1.4928946] [PMID: 26339305]

[91]   Desmaële D, Renaud L, Tingry S. A wireless sensor powered by a flexible stack of membraneless enzymatic biofuel cells. Sens Actuators B Chem 2015; 220: 583-9.
[http://dx.doi.org/10.1016/j.snb.2015.05.099]

[92]   Kotay SM, Das D. Biohydrogen as a renewable energy resource—prospects and potentials. Int J Hydrogen Energy 2008; 33(1): 258-63.
[http://dx.doi.org/10.1016/j.ijhydene.2007.07.031]

[93]   Ehsani M, Gao Y, Emadi A. Modern electric, hybrid electric, and fuel cell vehicles: fundamentals, theory, and design. CRC press 2017.
[http://dx.doi.org/10.1201/9781420054002]

[94]   Chorbadzhiyska E, Hubenova Y, Hristov G, Mitov M, Eds. Microbial electrolysis cells as innovative technology for hydrogen production. Proceedings of the Fourth International Scientific Conference FMNS-2011. 422-27.

[95]   Rozendal RA, Hamelers HV, Euverink GJ, Metz SJ, Buisman CJ. Principle and perspectives of hydrogen production through biocatalyzed electrolysis. Int J Hydrogen Energy 2006; 31(12): 1632-40.
[http://dx.doi.org/10.1016/j.ijhydene.2005.12.006]

[96]   Hu H, Fan Y, Liu H. Hydrogen production using single-chamber membrane-free microbial electrolysis cells. Water Res 2008; 42(15): 4172-8.
[http://dx.doi.org/10.1016/j.watres.2008.06.015] [PMID: 18718624]

[97]   Call D, Logan BE. Hydrogen production in a single chamber microbial electrolysis cell lacking a membrane. Environ Sci Technol 2008; 42(9): 3401-6.
[http://dx.doi.org/10.1021/es8001822] [PMID: 18522125]

[98]   Liu H, Hu H, Chignell J, Fan Y. Microbial electrolysis: novel technology for hydrogen production from biomass. Adv Biochem Eng Biotechnol 2010; 1(1): 129-42.
[http://dx.doi.org/10.4155/bfs.09.9]

[99]   Jeremiasse AW, Hamelers HV, Buisman CJ. Microbial electrolysis cell with a microbial biocathode. Bioelectrochemistry 2010; 78(1): 39-43.
[http://dx.doi.org/10.1016/j.bioelechem.2009.05.005] [PMID: 19523879]

[100]   Bandyopadhyay A, Stöckel J, Min H, Sherman LA, Pakrasi HB. High rates of photobiological H_2 production by a cyanobacterium under aerobic conditions. Nat Commun 2010; 1: 139.
[http://dx.doi.org/10.1038/ncomms1139] [PMID: 21266989]

[101]   Cheng S, Xing D, Call DF, Logan BE. Direct biological conversion of electrical current into methane by electromethanogenesis. Environ Sci Technol 2009; 43(10): 3953-8.
[http://dx.doi.org/10.1021/es803531g] [PMID: 19544913]

[102]   Cheng S, Logan BE. Sustainable and efficient biohydrogen production *via* electrohydrogenesis. Proc Natl Acad Sci USA 2007; 104(47): 18871-3.
[http://dx.doi.org/10.1073/pnas.0706379104] [PMID: 18000052]

[103]   Fan Y, Xu S, Schaller R, Jiao J, Chaplen F, Liu H. Nanoparticle decorated anodes for enhanced current generation in microbial electrochemical cells. Biosens Bioelectron 2011; 26(5): 1908-12.
[http://dx.doi.org/10.1016/j.bios.2010.05.006] [PMID: 20542420]

[104]   Freguia S, Rabaey K, Yuan Z, Keller J. Electron and carbon balances in microbial fuel cells reveal temporary bacterial storage behavior during electricity generation. Environ Sci Technol 2007; 41(8): 2915-21.
[http://dx.doi.org/10.1021/es062611i] [PMID: 17533858]

[105]   Hu H, Fan Y, Liu H. Hydrogen production in single-chamber tubular microbial electrolysis cells using non-precious-metal catalysts. Int J Hydrogen Energy 2009; 34(20): 8535-42.
[http://dx.doi.org/10.1016/j.ijhydene.2009.08.011]

[106]   Olivares-Ramirez J, Campos-Cornelio M, Godínez JU, Borja-Arco E, Castellanos R. Studies on the hydrogen evolution reaction on different stainless steels. Int J Hydrogen Energy 2007; 32(15): 3170-3.
[http://dx.doi.org/10.1016/j.ijhydene.2006.03.017]

[107]   Navarro-Flores E, Chong Z, Omanovic S. Characterization of Ni, NiMo, NiW and NiFe electroactive coatings as electrocatalysts for hydrogen evolution in an acidic medium. J Mol Catal Chem 2005; 226(2): 179-97.
[http://dx.doi.org/10.1016/j.molcata.2004.10.029]

[108]   Rozendal RA, Hamelers HV, Rabaey K, Keller J, Buisman CJ. Towards practical implementation of bioelectrochemical wastewater treatment. Trends Biotechnol 2008; 26(8): 450-9.
[http://dx.doi.org/10.1016/j.tibtech.2008.04.008] [PMID: 18585807]

CHAPTER 5

Nitrogen Removal in Bio Electrochemical Systems

Edris Hoseinzadeh[1], Hooshyar Hossini[2], Mahdi Farzadkia[3], Mahshid Loloei[4,*], Reza Shokuhi[5], Bahram Kamarei[6] and **Reza Barati Rashvanlou[3]**

[1] *Department of Environmental Health Engineering, Social Determinants of Health Research Center, Saveh University of Medical Sciences, Saveh, Iran*

[2] *Department of Environmental Health Engineering, Kermanshah University of Medical Sciences, Kermanshah, Iran*

[3] *Department of Environmental Health Engineering, School of Public Health, Iran University of Medical Sciences, Tehran, Iran*

[4] *Department of Environmental Health Engineering, Kerman University of Medical Sciences, Kerman, Iran*

[5] *Department of Environmental Health Engineering, Hamadan University of Medical Sciences, Hamadan, Iran*

[6] *Department of Environmental Health Engineering, School of Health and Nutrition, Lorestan University of Medical Sciences, Khorramabad, Iran*

Abstract: In recent years Bioelectrochemical systems (BESs) as a new approach represent an energy-efficient way for wastewater treatment, metal/nutrient recovery, and transformation of wastes to valuable products such as hydrogen. These unique characteristics of BESs are provided by a flexible platform for oxidation-reduction reactions. Although this process is still not fully developed but it can be nominated as a future trend of green energy and cleaner biochemical production pathway along with the waste remediation. Conversion of organic wastes into electricity and hydrogen/chemical products occurs in microbial fuel cells (MFCs) and microbial electrolysis cells (MECs), respectively. Different aspects of the process such as biological communities, physical configuration, and electrochemical conditions can affect a good performance of BES. In the present chapter review, the different fundamental mechanisms of BESs with emphasis on nitrogen removal have been reviewed. The behavior of affecting parameters on nitrogen and organic removal was discussed, including content pH, electrode material, level of supernatant concentration, the distance between electrodes and applied current, salinity and other operating conditions. Moreover, reported pathways and the theory of denitrification in BESs were described. Considering the finding, BES can be nominated as an alternative technology to the minimization of waste and a good way to generate electrical energy and valuable chemicals.

* **Corresponding author Mahshid Loloei:** Department of Environmental Health Engineering, Kerman University of Medical Sciences, Kerman, Iran; Tel: +9834313225103; E-mail: m_loloei@kmu.ac.ir

Keywords: Bioelectrochemical system, Denitrification, Electron donor, Electron acceptor, Microbial Fuel Cell, Nitrate, Nitrogen, Wastewater.

INTRODUCTION

In bio-electrochemical systems(BESs), microorganisms are used to catalyze electrochemical reactions through interaction with electrodes [1]. Typically, these microorganisms attach to the surface of the electrode for transferring electrons (donating or accepting) to solid electrodes, as well as stimulating microbial metabolism [2, 3]. The electrodes are also known as bio-electrode in BESs. It is possible to use BESs for treating sewage and producing renewable hydrogen for the generation of electricity [4] that some reactor schematics of used BESs showed in Fig. (**1**).

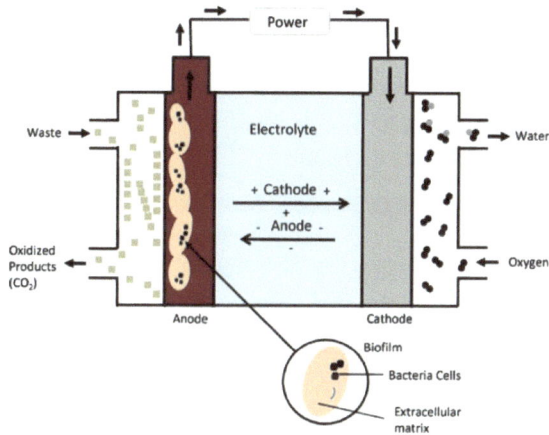

Biofilm of cathode reactor electrode.

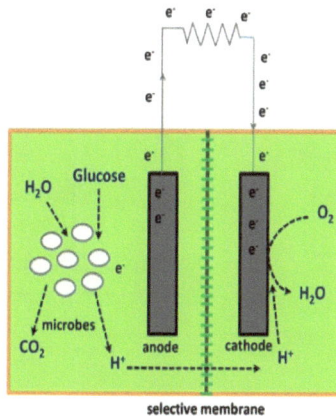

Fig. (1). Schematic of microbial fuel cells mechanism [5].

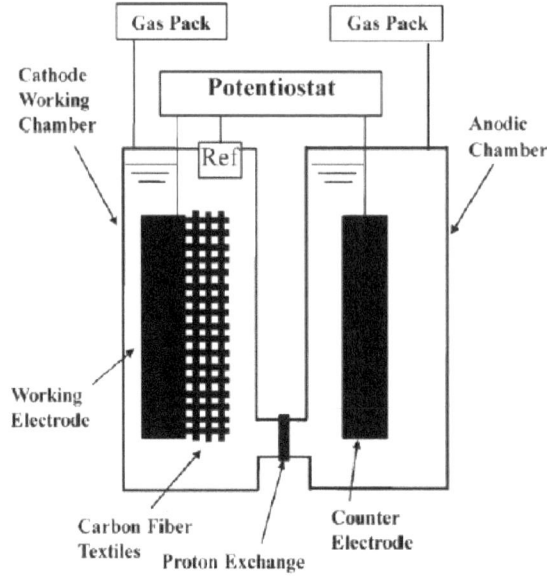

Biofilm of cathode reactor electrode.

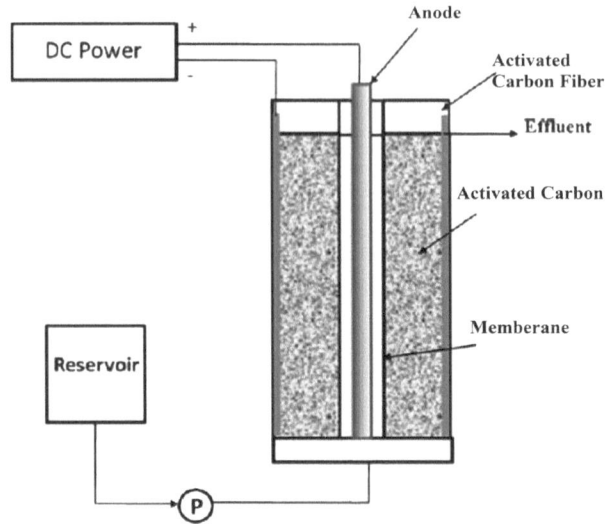

3D reactor for nitrate removal

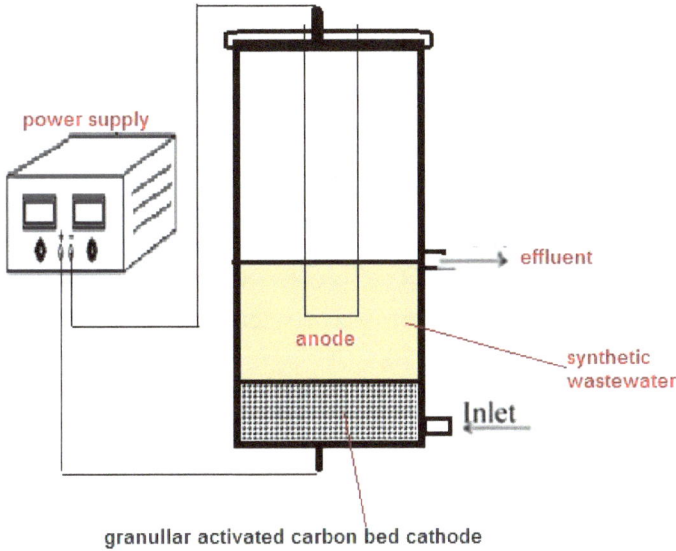

Fig. (2). Single-chamber bio-electrochemical system for treating nitrate infused water [5].

Since in BESs, wastes are converted into chemical energy, which is not harmful to the environment, hence they have the potential to be among the green technologies. BESs are divided into two main groups of microbial fuel cells (MFCs) and microbial electrolysis cells (MECs) [1, 5]. The amount of energy required to transfer electrons is expressed by the standard oxidation-reduction potentials (E°). If thermodynamic conditions for the redox reaction (a combination of reduction and oxidation) are provided, energy is produced, and the system operates as an MFC or *vice versa*. To use hydrogenotrophic denitrifying bacteria in the removal of nitrate, external electrical energy is required to produce hydrogen, because these types of microorganisms use hydrogen as a source of energy. Table **1** presents the common compounds used in BESs with their standard values.

MICROBIAL FUEL CELLS (MFCs)

The MFCs are anaerobiosis and produce electricity from biodegradable organic material. When sewage is used as the MFC substrate and electricity is generated, (Fig. **2**) the total cost of wastewater treatment will be low. A conventional MFC has two parts, consisting of an anodic chamber and a cathodic chamber, which are connected to each other by a proton exchange membrane (PEM) or salt bridge. This membrane or salt bridge allows protons to pass and enter the cathode chamber and can also prevent the release of oxygen from the anodic chamber. Despite all the advantages of MFC, the two- chambers MFC has a complex design and its real implementation is difficult. The single-chamber MFC has a simpler design, with a higher volumetric force density and lower cost [6]. The separator or membrane plays a critical role in the MFCs and can prevent the transfer of the proton from the anodic chamber to the cathode chamber. Therefore, a higher pH in the cathode chamber than the anodic chamber is more common and causes to reduce the stability of the system and its performance. In addition, the overall internal resistance of the MFC is decreased due to the membrane. Without the presence of the membrane, the release of oxygen and substrate is increased and resulted in a lower amount of coulombic efficiency and bio-electro catalytic activity in the anode [7, 8].

Table 1. Common compounds used in BESs as substrates.

Standard potential E° (V) (*vs.* NHE)	Anode Oxidation Reaction	Compound
0.187	$C_2H_3O_2^- + 4H_2O \rightarrow 2HCO_3^- + 9H^+ + 8e^-$	Acetate
0.104	$C_6H_{12}O_6 + 12H_2O \rightarrow 6HCO_3^- + 30H^+ + 24e^-$	Glucose
0.118	$C_3H_8O_3 + 6H_2O \rightarrow 3HCO_3 + 17H^+ + 14e^-$	Glycerol
0.01	$C_4H_5O_5 + 7H_2O \rightarrow 4H_2CO_3 + 11H^+ + 12e^-$	Malate
0.022	$C_6H_5O_7^{3-} + 11H_2O \rightarrow 6H_2CO_3 + 15H^+ + 18e^-$	Citrate
0.207	$C + 2H_2O \rightarrow CO_2 + 4H^+ + 4e^-$	Carbon
0.83	$NO_3^- + 2H^+ + 2e^- \rightarrow NO_2^- + H_2O$	Nitrate to nitrite
1.246	$2NO_3^- + 12H^+ + 10e^- \rightarrow N_2 + 6H_2O$	Nitrate to nitrogen gas
0	$2H^+ + 2e^- \rightarrow H_2$	Proton to hydrogen
1.229	$O_2 + 4H^+ + 4e^- \rightarrow 2H_2O$	Oxygen to water
0.694	$O_2 + 2H^+ + 2e^- \rightarrow H_2O_2$	Oxygen to hydrogen peroxide

Oxidation-Reaction-Potential (ORP) Reactions in MFCs

In MFCs, electricity is generated as a result of oxidation-reduction reactions, which result in electron release, transfer and acceptance through biochemical and electrochemical reactions in electrodes of the anodic and cathode chamber that one of electrode acts as an electron donating and the other one acts as the electron acceptor. The chemical compounds responsible for electron reception are called the terminal electron acceptor (TEA). The following (Table **2**) shows oxidation-reduction reactions that describe potential bioelectrochemical reactions in MFCs used to generate electricity in which sewage is treated as a substrate (electron donating agent) and other pollutants such as nitrate, phosphate, *etc.*, act as an electron acceptor [7, 8]:

Table 2. Oxidation-reduction reactions of potential bioelectrochemical reactions in microbial fuel cells [5, 9].

Oxidation reactions (anode)	
Potential	**Reactions**
Glucose: $E° = -0.429V$ *vs.* SHE	$C_6H_{12}O_6 + 12H_2O \rightarrow 6HCO_3^- + 30H^+ + 24e^-$
Glycerol: $E° = -0.289V$ *vs.* SHE	$C_3H_8O_3 + 6H_2O \rightarrow 3HCO_3^- + 17H^+ + 14e^-$
Malate: $E° = -0.289V$ *vs.* SHE	$C_4H_5O5- + 7H_2O \rightarrow 4H_2CO_3 + 11H^+ + 12e^-$
Sulfur: $E° = -0.230V$ *vs.* SHE	$HS^- \rightarrow S^0 + H^+ + 2e^-$
Reduction reactions (cathode)	
Reactions	Potential
$O_2 + 4H^+ + 4e^- \rightarrow 2H_2O$	$E° = + 1.230$ V *vs.* SHE
$O_2 + 2H^+ + 2e^- \rightarrow 2H_2O$	$E° = + 0.269$ V *vs.* SHE
$NO_3^- + 2e^- + 2H^+ \rightarrow NO_2^- + H_2O$	$E° = + 0.433$ V *vs.* SHE
$NO_2^- + e^- + 2H^+ \rightarrow NO + H_2O$	$E° = + 0.350$ V *vs.* SHE
$NO + e^- + H^+ \rightarrow 1/2N_2O + 1/2H_2O$	$E° = + 1.175$ V *vs.* SHE
$1/2N_2O + e^- + H^+ \rightarrow 1/2N_2 + 1/2H_2O$	$E° = + 1.355$ V *vs.* SHE
$2NO3- + 10e^- + 12H^+ \rightarrow N_2 + 6H_2O$	$E° = + 0.734$ V *vs.* SHE
$Fe^{3+} + e^- + H^+ \rightarrow Fe^{2+} + 1/2H_2O$	$E° = + 0.773$ V *vs.* SHE
$MnO_2 + 3e^- + 4H^+ \rightarrow Mn^{2+} + 2H_2O$	$E° = + 0.602$ V *vs.* SHE

Denitrification in MFCs

In biological denitrification, nitrate can receive electrons from organic compounds and reduces it to nitrogen gas (for example, in conventional biological denitrification). The electron transfer process allows the use of nitrate as the final electron acceptor in the BES. In the case of using an organic compound (such as acetate) as an electron source, the reduction of nitrate can produce a positive electrical potential of 0.98 V. In an MFC, if bacteria directly use a cathode electrode as an electron donor, the electricity needed to provide the reductant power required for denitrification can be significantly reduced. Nitrogen can affect the performance of BES, in particular, can have an effect on the generation of electricity through the inhibitory effect on germs, adjusting pH and competition for the electron donor/receptor. According to the available reports, total ammonia nitric (TAN) concentrations above 500 mg/L can have a significant inhibitory effect on the generation of electrical current and the maximum amount of electricity production is decreased from 4.2 to 1.7 W/m^3 by increasing the concentration of TAN from 500 to 4000 mg/L [10]. It can be said that high concentrations of free ammonia have inhibitory effects on anodic respiration bacteria (ARB). One of the main factors affecting nitrogen removal performance in MFCs is dissolved oxygen (DO), pH, carbon to nitrogen ratio, and electricity generation through anodic process. A high amount of DO and pH prevent the denitrification process. Neutral pH is suitable for nitrification-denitrification processes in MFCs [10]. The cathodic process is affected by a number of electrons released through the anodic process, so the ratio of carbon to nitrogen is an important parameter for process performance. It seems the high C/N ratio is better for denitrification. However, higher C/N ratios, due to the dominance of heterotrophic denitrification, may have an effect on MFC performance.

Microbial Electrolysis Cell (MEC)

Microbial electrolysis cell (MEC) is used to treat wastewater due to the decomposition of organic matter and the conversion of sewage chemical energy to hydrogen gas by applying electrical energy. Exoelectrogenic microorganisms are used to oxidize organic matter and produce hydrogen, carbon dioxide and electrons. Anode acts as an electron acceptor and then transmits electrons through an external circuit to the cathode; while H^+ in the cathode is reduced to molecular hydrogen by applying the current. The main advantage of this system is its low energy consumption, so that it requires 2.0-14.0 V energy to produce hydrogen continuously by electrohydrogenisis (or hydrogen production from organic matter decomposition by bacteria), while the minimum theoretical electrical energy required for electrolysis of water is about 23.1 V. MEC is more advanced than other types of biological processes of hydrogen production, so that it is capable of decomposing many fermentable and non-fermentable organic substances, and its hydrogen production efficiency is high [11 - 13].

Bioelectrodes

Bio-electrodes are bio-anodes or bio-cathodes; in an anodic chamber, microorganisms use existing substrates as carbon sources and electron donors to produce the energy carrier molecule known as adenosine-5-triphosphate (ATP). Organic matter is converted into citric acid by glycolysis, which is ultimately oxidized to carbon dioxide. Nicotinamide adenine dinucleotide (NAD$^+$) and flavin adenine dinucleotide (FAD) are also reduced into their electron carriers, NADH and FADH$_2$. These electron carriers transfer their electrons from the cytoplasm (the place of the citric acid cycle) to the cell membrane and then move towards the anode through the direct electron transfer mechanisms or through the mediators [5, 13]. In other words, anode plays an important role as an external electron acceptor for oxidation of the organic substrate (Fig. **3**). In bio-cathodes, bacteria are used as biocatalysts to receive electrons from electrodes to replace chemical catalysts. The electrons must pass through electron receptors with high electro positivity such as oxygen, nitrate, and chlorine organic compounds by cytochromes of the external membrane through the inner membrane and the periplasm, so the standard oxidation/reduction potential (E°) tends to be higher, indicating electrons are readily absorbed by microorganisms and are sufficient to store energy.

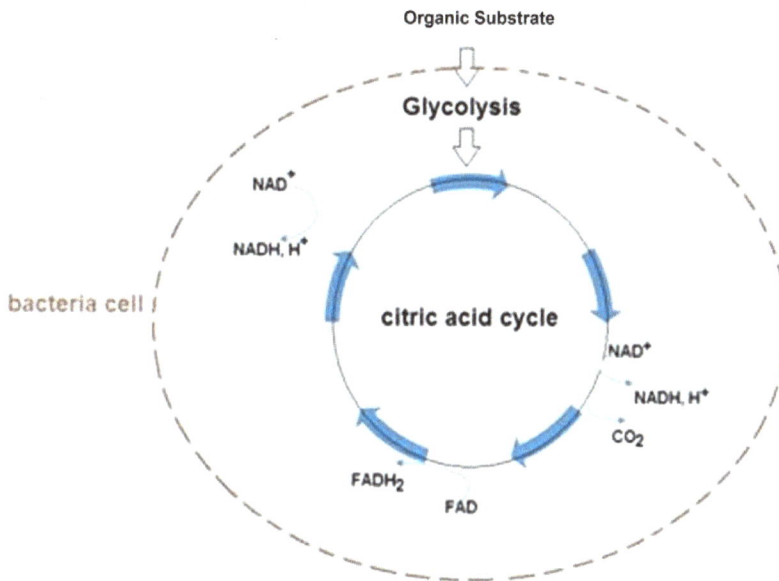

Organic Substrate

Glycolysis

NAD^+

$NADH, H^+$

bacteria cell

citric acid cycle

NAD^+

$NADH, H^+$

CO_2

$FADH_2$

FAD

Fig. (3). Reduction of NAD $^+$ and FAD through a citric acid cycle.

Direct Electron Transfer (DET)

DET is defined as direct contact between the bacterial enzymes derived from the central cell membrane (internal membrane, type C cytochrome and periplasm), and the surface of the electrode (Fig. **4**). These microorganisms rely on an electron transfer protein bound to the membrane for transferring electrons from the bacterial cell to its outer surface (electrode) or *vice versa*. Some researchers have concluded that the direct transfer of electrons is very slow due to the location of the active site of the enzyme that is located within the protein environment. Some of the redox enzymes of exoelectrogenic bacteria are located on the outer surface of the membrane of microorganisms, so active sites of redox enzymes are directly in contact with electrodes or the environment. The DET method requires physical contact between bacterial cells, cytochrome and electrode. Of course, only bacteria in contact with the electrode are electrochemically active. The bacteria used in anode and cathode reactions are located in Shewanella, Rhodoferax and Geobacter families. Some Geobacter and Shewanella strains are able to produce pili or conductive nanowires that help bacteria connect to the electrode. The nanowires are attached to the membrane cytoplasm and help to transfer electrons inside or outside the cell. As nanowires can keep microorganisms even at intervals of 40-50 microns from the electrode surface continuously, so multilayer electro-active biofilms can also be formed on the electrodes, that the overall result of which is increasing the performance of bio-electrodes [14, 15].

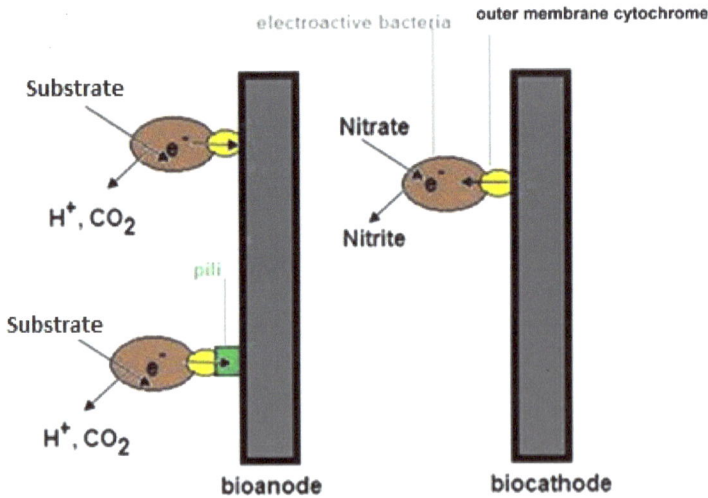

Fig. (4). Direct electron transfer mechanisms in bio-electrodes: a) Bio-anodes: i) Electron transfer by cytochrome of the external membrane; ii) Electron transfer by nanowires. b) Bio-cathodes [5].

Mediated Electron Transfer

Since in most microorganisms, the outer layers are formed of nonconductor lipid membranes, peptidoglycans and lipopolysaccharides, thus they can transfer electrons to electrodes directly. Therefore, an electron mediator needs to transfer electrons between electrodes and microorganisms. Common synthetic exogenous mediators include methyl viologen, anthraquinone-2,6-disulfonate (AQDS), neutral red, humic acids and sulfur. These redox mediators are not used by microorganisms and may be recovered at the electrode surface. In Table **3**, some commonly used electron conductors with their redox potential are presented. The use of exogenous redox mediators to stimulate electron transfer in BES is not an environmentally friendly process and their use can be considered as a risk factor for health. Another type of MET mediator does not require synthetic redox mediators and will be produced by themselves. Some microorganisms are capable of synthesizing redox mediators through primary and secondary metabolites that result in electron transport and are independent of exogenous electron conductor transducers. Mediators act as a reversible electron acceptor or donor and transfer electrons from bacterial cells to the anode or from the cathode to bacterial cells. Among bacterial-mediated producers can be pointed to Phenazine (a redox mediator) produced by *Pseudomonas aeruginosa* and it has been used in transferring electrons between bacteria and anode [5, 16, 17].

Table 3. Some common electron mediators with their redox potential [5, 9].

Redox Potential E°′ (V)	Mediator Redox
-0.29	Safranin
0.02	Gallocyanin
-0.32	Neutral Red
-0.18	Anthraquinone -2,6-Disulfonate (AQDS)
-0.45	Methyl Viologen
0.79	Iron

FACTORS CONTROLLING REMOVAL OF ORGANIC MATTER IN BIO-ANODES

The biofilm on the anode surface will hydrolyze the complex organic matter into simpler molecules before being electrochemically oxidized by active germs. Since the food industry wastewater contains organic materials that is simply oxidized by microorganisms, its treatment with BES is successful. Microorganisms in the anode chamber use electrons and protons under anaerobic conditions are used to decompose organic matter for preventing the consumption of electrons by oxygen and the formation of water. These electrons are transferred from the organic electron donors to the anode by bacterial respiration enzymes. The feature of microorganisms used in BES is the ability of anaerobic hydrolyze of cellulose, good electrochemical activity and the use of anode as an electron acceptor during the oxidation of cellulose metabolites by hydrolysis. There are some factors that play an important role in the bioavailability of organic matter, which are discussed in the following.

Effect of pH

pH is an important parameter in control of the activity of ARB. The optimum pH is between 6 and 7 for the microbial activity. Higher removal of COD occurs at pH values higher than 7, but it is noteworthy that the carbon used in the methanogenic process, leads to the formation of methane due to microbial activity and is a greenhouse gas. Bacteria gain energy through electron transfer from electron to an electron acceptor. The process of donating electrons leads to the production of electrons consumed by electron acceptors and, are known as the half-reaction of oxidation and the half-reaction of reduction, respectively. In the microbial processes, both half-reactions are combined and form an equilibrium reaction. In BESs, these two half-reactions occur in separate locations, namely, the electron donating half-reaction (oxidation) in the anode and the reduction half-reaction in the cathode. Although anode has a role in accepting an electron, but it only transfers electrons to complete the circuit without changing the oxidation state of electro-active species. In this case, anodic oxidation of organic matter leads to the production of H^+ ions, which results in a decrease in the pH of the anodic chamber. Therefore, this decrease in the pH reduces significantly the performance of ARBs compared to their maximum activity, which occurs at neutral pH. To solve this problem, a buffer (phosphate buffer is commonly used.) or carbonate is used that is combined with H^+ and forms a weaker acid [18, 19]. The following relationships show the equilibrium reactions of acid-base and acid-carbonate buffers, that in which Alk⁻ refers to alkalinity, HAlk is protonated alkalinity, CO_3^{2-} is carbonate ions, HCO_3^- is bicarbonate ions, H_2CO_3 is carbonic acid, and CO_2 is carbon dioxide gas [5].

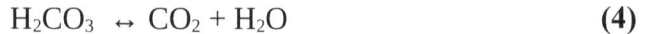

$$Alk^- + H^+ \leftrightarrow HAlk \tag{1}$$

$$CO_3^{2-} + H^+ \leftrightarrow HCO_3^- \tag{2}$$

$$HCO_3^- + H^+ \leftrightarrow H_2CO_3 \tag{3}$$

$$H_2CO_3 \leftrightarrow CO_2 + H_2O \tag{4}$$

Some researchers in their studies concluded that with the presence of the total buffer in solution, the density of the current generated in the BES is increased, which is because of expansion of the active biofilm to a greater depth of the expected layers and due to the buffer state, which prevents pH changes and so, the effect on the bacterial function. The use of carbonate buffer is better than phosphate buffer in pH adjustment because mineral carbon is present in all natural waters and has a higher diffusion coefficient in water, resulting in faster movement of HAlk from the biofilm. In addition, carbonate and bicarbonate can be reused in the control of pH because carbon dioxide gas is produced through reactions of acid-carbonate buffer and recovered internally in a cathode and then reacts with oxygen to produce carbonate and bicarbonate. Carbonate and bicarbonate ions released into an anode chamber again, and this cycle regenerated.

ANODE MATERIAL AND ITS AVAILABLE SURFACE

Although platinum anodes have been successfully used to remove organic pollutants, the use of this type of anode can lead to a high cost of the initial investment in the process. Stainless steel is one of the most used materials in making an anode, and various studies have reported good results in terms of reducing the amount of organic pollutants. However, the use of carbon-based anodes, such as graphite and carbon, has been recommended by many researchers because of sustainability, especially when biofilms are formed on them and so, the overall cost of the process is cheaper. Pretreatment of carbon electrodes with ammonia and oxidation in sulfuric acid or nitric acid is an essential step in improving the microbial population of the biofilm formed on it and improving the electron transfer because, with this kind of pretreatment, carboxylic groups will be created on the carbon electrode. Bacteria that are able to catalyze the oxidation of organic matter and direct electron transfer can easily form graphite colony on electrodes and can easily transfer electrons. Despite the positive features of these types of electrodes, the ohmic resistance of carbonate anodes is almost 1000 times greater than metal ones [2, 3, 5]. In some studies, the coating of graphite electrodes using electron mediators, active polymers, polyaniline (PANI) and quinone groups, has been used as methods for modifying and improving the performance of carbonic anodes. Among metal electrodes, the use of copper electrodes is not recommended due to the probable production of toxic species as a result of copper dissolution and is classified as inappropriate electrodes. The higher available surface for the electrode provides more space for the bacteria to attach the electrode, and thus the electron transfer rate/amount is increased. Of course, the result mentioned is true for the condition in which the resulting force and electricity generation depends on biofilm attached to the electrode surface.

DISTANCE BETWEEN ELECTRODES

Internal resistance, also called ohmic loss, is one of the main limitations of BESs and should be consider calculating additional potential, which means compensating the potential less over the electron and proton transfer process. The internal resistance can be reduced by reducing the space between the electrodes. The electrodes are very close in the single chamber MFCs, and as oxygen is released from the cathode chamber to the anode chamber, it can reduce the performance. The distance between the electrodes can be very low for MECs, because this type of system typically use under anaerobic conditions.

CONCENTRATION OF ORGANIC MATTER

The concentration of organic matter (substrate) not only affects the size of the bacteria and the biofilm morphology, but also affects the density of the electrical current produced and the coulombic efficiency of BES. Although the higher initial concentration of organic matter stimulates the removal of organic matter, it also results in a decrease in coulombic efficiency because organic matter is consumed by other organic processes, such as fermentation and methanogenesis, and is converted into lateral products of these processes.

BIO-ANODE PERFORMANCE

It is believed that the microorganisms in the anode chamber act as biocatalysts, convert the energy stored in the chemical bonding of the substrate to electrical energy, and transfer the electrons from the substrate oxidation to organic electron receptors, and bioelectricity will be produced. In addition, the pH of the anode chamber increased because of microorganism activity (removal of COD). In other words, the pH of the anode chamber affect by the removal of COD and substrate.

DENITRIFICATION CONTROLLER FACTORS IN BIO-CATHODE

As there is no need to remove excess substrates that are available to facilitate the production of biomass in autotrophic denitrification compared to heterotrophic type, autotrophic denitrification has higher overall efficiency. Generally, in autotrophic denitrification, hydrogen gas consider as an electron donor because it has a lower supply cost and does not produce any toxic by-products. Of course, hydrogen gas has low solubility in water and has the explosive capability. Accordingly, many researchers have proposed the stabilization of denitrifying bacteria on the cathode in the BES system that denitrifying bacteria use the hydrogen produced by electrolysis of water as the required hydrogen. In other words, during the microbial denitrification, the cathode acts as an electron donor. The denitrification reactions are used hydrogen produced by electrolysis of water in the cathode (according to the following relationships) [5].

$$2H_2O + 2e \rightarrow H_2 + 2OH^- \tag{5}$$

$$NO_3^- + H_2 \rightarrow NO_2^- + H_2O \tag{6}$$

$$2\,NO_2^- + 3H_2 + 2H^+ \rightarrow N_2 + 4H_2O \tag{7}$$

$$\text{General reaction } 2NO_3^- + 5H_2 + 2H^+ \rightarrow N_2 + 6H_2O \tag{8}$$

pH Effect

Sewage pH is the main factor affecting the performance of hydrogenotrophic denitrification process. When the pH value is greater than 8.8, nitrite concentration is higher and at pH less than 7, carbonate ions are decomposed and results in a reduction in the removal rate of nitrate [20]. During the batch decontamination process, the pH of the solution is increased that is usually adjusted between 6.5 and 7 using phosphoric acid. Moreover, the pH adjustment occurs because of the carbon dioxide production in the anode that controlled by the electric current [1, 9]. The following relationships show the mechanism of pH adjustment by carbon dioxide dissolved in water [21, 22]. In this case, by dissolving carbon dioxide in water, carbonic acid is produced that reacts with OH⁻ and bicarbonate is produced. Then the bicarbonate reacts with OH-ions and carbonate is formed. This mechanism shows that carbon dioxide can increase the electrical conductivity of water and, since more ions (carbonate and bicarbonate) are dissolved in water, it can also reduce the ohmic potential of the omega. Some studies have shown that without pH adjustment during the process, the amount of nitrate removal is low (about 26%), while the nitrate removal rate has been increased by adjusting and keeping pH level within neutral pH values. In general, the optimal pH for biological denitrification has been reported in the range of 6.5-8.

$$CO_2 + H_2O \rightarrow H_2CO_3 \tag{9}$$

$$H_2CO_3 + OH^- \rightarrow H_2O + HCO_3^- \tag{10}$$

$$HCO_3^- + OH^- \rightarrow H_2O + CO_3^{-2} \tag{11}$$

Electric Current Effect

The applied current is a major factor affecting the rate of hydrogen produced in the cathode chamber and indirectly plays a vital role in the regeneration of nitrate [4, 13]. The autotrophic denitrification is directly proportional to the current and indicates that BES effectively uses electrons through affecting the speed of nitrogen gas production. Researchers have shown that bio-cathodic denitrification is better done in a lower electrical current because of the lower production of hydrogen gas, which tends to restrict the process. By applying a higher electrical current, hydrogen gas production increased through electrolysis of water. Hydrogen gas bubbles result in the formation of channels in the granular carbon activated (GAC) bed, and form a dry atmosphere, and ultimately lead to a decrease in the performance of denitrification. The lack of full use of hydrogen gas also reduces the function of denitrification. On the other hand, applying a higher electrical current intensifies the production of oxygen in the anode, which reduces production of oxygen over the hydrogenotrophic denitrification reaction. A higher electrical current density leads to an increase in the rate of denitrification, but reduces the efficiency of denitrification-current (η). Because denitrification in BES involves the application of current, accurate estimation of nitrate reduction during the process can be obtained from the efficiency of denitrification [10, 19, 23]:

$$\eta = \frac{Q(C_{NO_{B,i}^-} - C_{NO_{B,f}^-})}{I/nF} \qquad (12)$$

In this equation, Q is the volumetric flow rate (cm^3/s), $C_{NO_{B,i}}$ and $C_{NO_{B,f}}$ are nitrate concentration (mol/cm^3) in the input and output of the system, n is the stoichiometry coefficient (n= 5) and F is Faraday constant (C/mol). Some researchers have proposed the effect of current and hydraulic retention time on the removal rate of nitrate with terms of "current efficiency" (CE). CE is defined as the ratio of the electric current consumed in producing a target product to the total electric current consumed. CE is calculated using the following equation [5]:

$$CE = [(C_{NO_3^-,\epsilon} - C_{NO_3^-,eff}) \times 5 + (C_{NO_2^-,\epsilon} - C_{NO_2^-,eff}) \times 3]$$
$$\frac{V}{HRT} \times F \times \qquad (13)$$

In this equation, C is the concentration of nitrate or nitrite (mmol-N/L), F is Faraday constant (26.8 mAh/mmol), and V is the effective volume of the reactor (L). The term of "flow efficiency" shows the specificity of the process as well as the electro catalytic function associated with surface reaction and mass transfer in the system. The relationship between CE, flow and HRT in different studies shows that HRT and unwanted current due to excess hydrogen produced without any increase in pH, insufficient contact time between a substrate and biomass and increasing pH to more than its optimal range, reduce CE.

The Electrodes Material Effect

Carbon materials have a suitable mechanical power and roughness surface, which causes carbon nanoparticles preferable to carbon nanotubes for forming biofilms. Of course, the use of carbon materials is difficult to apply to large-scale processes due to their brittleness and bulky nature, as well as their high electrical resistance, which increases the dielectric loss of the electrodes. In many cases, conductive metal supports such as stainless steel mesh are used for graphite and carbon electrodes. Many researchers have suggested using acetyl and platinum in many cathodes because of their better mechanical, control, and electro-kinetic properties compared to carbon. In platinum electrodes, the formation of a platinum oxide layer (PtO) on the electrode surface is likely that formation of these layer interrupts the process of denitrification. The use of graphite granules has been proposed in many studies due to their high available surface, which facilitates sticking of bacteria and consequently, biofilms formation (that in BESs, can be used as a third bipolar electrode).

Bio-Cathode Performance

Studies have shown that the use of electrodes of various materials can be suitable as a substrate for the formation of microbial biofilm, but when some electrodes like copper are used, its dissolution and the presence of heavy metal ions prevent the process of denitrification to continue. Moreover, some electrodes affect the process through their conductivity. Because the speed of denitrification is directly proportional to the electric current that controls the electrolysis of water and is considered as the speed limiting step.

REMOVAL OF ORGANIC COMPOUNDS SIMULTANEOUS WITH DENITRIFICATION

Systems used to simultaneous removal of organic matter and nitrates consist of two anodic and cathode chambers separated by a proton exchange membrane. The electrons and protons that are produced from the anodic oxidation of organic matter are transferred to the cathode chamber through salt bridge and membrane, respectively. Electrons are used to complete the denitrification process, while protons are used to regulate cathodic pH. Studies have reported that the use of MFC is appropriate for the treatment of organic matter and low concentrations of ammonia [24]. This type of wastewater is generally produced in food processing industries such as breweries, starch processing, and so on. In some studies, sewage input to the BESs with a C/N ratio of less than 2.8 reported to be suitable for the simultaneous removal of COD and nitrate. The results of some studies show that cathodic and anodic chambers should be located under anaerobic conditions in order to limit the oxygen leakage into the anodic chamber and, as a result, to increase the efficiency of the anodic reaction.

REMOVAL OF NITROGENOUS COMPOUNDS IN BIO-ELECTROCHEMICAL SYSTEMS (BESs)

BESs are a clean technique with the ability to recover energy from sewage. The BES is based on electrochemical active microorganisms with the ability to exchange electrons between electrodes and organic matter for converting chemical energy to electrical energy or other types of energy. The electrochemical cells used by microorganisms to catalyze oxidation or reduction reactions in electrodes (anode or cathode) are generally called as BES. Typically, bio-anaerobic bacteria oxidize organic matter to carbon dioxide in the sewage using anode as the terminal acceptor of the electron. The oxidation reaction in the anode can be expressed by decomposition of acetate or glucose as shown below [7, 10]:

$$CH_3COO^- + 4H_2O \rightarrow 2HCO_3^- + 9H^+ + 8e^- \tag{14}$$

$$C_6H_{12}O_6 + 12H_2O \rightarrow 6HCO_3^- + 30H^+ + 24e^- \tag{15}$$

Anaerobic respiration of microorganisms releases electrons from organic matter that are in a higher potential of organic matter and they transfer to the anode. Microorganisms use external cellular electron transfer mechanisms (such as direct transfer, electron shuttle, biofilm matrix or pili) to transfer electrons to a solid electron acceptor (*i.e.*, anode) located outside the cells. In this method, anaerobic bacteria store the energy needed for their growth. The electrons are then transferred to the cathode through an external circuit (and resistor). Ultimately, electrons are used to reduce electron acceptors such as oxygen, nitrate, and metal ions in the cathode. In addition to the electron, the proton is also released due to organic matter decomposition in the anode chamber, and hydroxyl ions (OH⁻ ions) are produced in cathode reactions. The accumulation of protons in the anode chamber leads to its acidification and, as a result, the metabolic activity of electro-biofilms is decreased. For each electron transferred through an external circuit, one proton should be transferred to the cathode chamber, and the proton is consumed by the hydroxyl ions produced in the cathode. Various types of energy conversion have led to research in many areas, including generating electricity from sewage, producing various types of chemicals, converting electricity to methane or organic compounds. In some studies, BES has been used to remove nitrogen compounds and generate electricity.

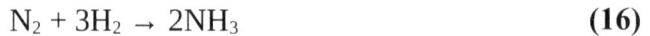

$$N_2 + 3H_2 \rightarrow 2NH_3 \tag{16}$$

The hydrogen gas required for the Haber-Bosch process is produced through natural gas deformation. Most ammonia produced industrially (approximately 80 Mt/y) are used as fertilizers in farms. A part of the ammonia is consumed by the plant, and a part of it enters into the environment, and increases the risk of the eutrophication phenomenon in the acceptor water resource, and also a part can be introduced into the atmosphere. In water resources, ammonium is generally in the form of nitrogen, and in sewage treatment plants, removal of ammonium is carried out through a two-stage process. Firstly, nitrification is carried out under

Fig. (5). Removal of nitrogen compounds in bio-electrochemical systems: 1) Ammonia transfer through the membrane: 1a) passively through the release of ammonia or 1b) through active displacement in the form of ammonium; 1c) exiting ammonia from the system by increasing the pH in the cathode; 2a) biological oxidation in the cathode chamber by oxygen and denitrification by cathode-attached microorganisms; or 2b) microorganisms suspended in solution (discontinuous lines indicate the process independent of the electrodes); 3) direct removal of ammonia by microorganisms through nitrification/denitrification, 4) consuming ammonium in microorganisms growth.

anaerobic conditions and ammonia is oxidized to nitrite by Nitrosomonas, and next, nitrite is oxidized to nitrate by nitrobacterium. For full oxidation of ammonium to nitrate, two moles of oxygen are needed for each mole of ammonium. In the second step of the biological removal of ammonium, the nitrate produced in the previous stage converted to nitrogen gas by *paracoccus denitrificans* under the anoxic condition that is either, recovered or released into the atmosphere. The disadvantage of the common nitrification/denitrification reaction is requiring significant amounts of energy source because aeration is needed to supply the oxygen needed for converting ammonium to nitrate and COD (for example, methanol) is needed to supply the electrons required for denitrification. However, the advantage of the nitrification/denitrification process is that it can remove nitrogen in low concentrations. Other improvements performed on the biological processes for the removal of nitrogen compounds have been described in the relevant section and reintroducing them is avoided. BESs are one of the improvements performed on biological systems for the biological elimination of nitrogen compounds that has many advantages over biological types. In BESs, a series of reactions occurring both in the anode and in the cathode will lead to the removal of organic matter or nitrogen compounds (Fig. **5**).

The first mechanism for the removal of ammonium is based on an active or passive transfer through an ion exchange membrane, which combines with acid/base equilibrium [5, 10]. Ammonium concentration in sewage is considerable, for example, >4 g/L in urine and 0.04 g/L in household sewage; consequently, in sewage with high concentration of ammonium, the major ion transferred through the membrane (either as a passive (In the form of NH_3 without charge) or active form (in the form of NH_4^+ with electric charge) is ammonium. In catholyte, ammonium is released in the form of ammonia due to the increase in pH in the cathode chamber that changes the chemical equilibrium from ammonium to ammonia. The second proposed mechanism is the reduction of nitrate to nitrogen gas (denitrification) by the microorganisms available in the cathode. In this mechanism, nitrates must first be formed (for example, through biological oxidation of ammonium and under anaerobic conditions). There is another possible mechanism that involves converting ammonium into nitrogen gas through the anammox process; this process is essentially independent of cathode processes. The third mechanism that may occur in the anode is that ammonium is converted directly to nitrogen by microorganisms, although this mechanism has still not been verified scientifically. The fourth mechanism is the use of ammonium in the biomass during microbial growth in an anode or cathode. In the case of the oxidation of ammonium to nitrogen gas in the anode, the results of studies have shown that ammonium use to generate electricity directly as anodic fuel or otherwise indirectly as a substrate needed for nitrification to produce organic compounds for heterotrophs and so, it is eliminated. Moreover, the inverse of this result has been reported, that instead of biological elimination of ammonium (oxidation to nitrogen gas in the anode), it is eliminated in the cathode due to physicochemical factors such as releasing through the membrane and evaporation due to an increase in pH. However, some researchers have rejected the direct oxidation of ammonia in the anode and its use as fuel for electricity production. Moreover, cyclic voltammetry experiments did not show redox pairs, nitrites, nitrates and nitrifying or anammox species have not been identified in the anode. In general, considering the results of available and sometimes contradictory reports, it can be concluded that direct oxidation of ammonium into nitrogen gas cannot be verified.

NITRIFICATION/DENITRIFICATION IN CATHODE

Some researchers reported full-length denitrification by microorganisms in the cathode of MFC [25, 26]. The acetate/organic matter is consumed as the electron in the anode and the nitrate is used as the electron acceptor in the cathode. Therefore, the generation of electricity by nitrifying microorganisms and the reduction of nitrate to nitrogen gas simultaneously occurred. According to available reports, the nitrate removal rate in such systems estimated about 146 gN/m^3d. In these systems, if the input wastewater of the system contains ammonium, the generation of electricity over ammonium oxidation is likely. In these systems, synchronous nitrification-denitrification can be achieved by optimizing the oxygen supply in the cathode.

BIO-ELECTROCHEMICAL DENITRIFICATION IN BIOFILM ELECTRODE REACTORS (BERs)

A biological denitrification using electricity is theoretically performed through the production of hydrogen at the cathode surface (not at the anode surface). Hydrogen and low oxidation-reduction potential of the environment are created through cathode reactions that can be used by hydrogenotrophs to reduce nitrate to nitrogen gas. Proper contact between microorganisms and hydrogen will lead to improve biological denitrification. BERs include the direct fixation of the denitrifying bacteria on the cathode surface, where hydrogen is produced, thereby facilitating the access of microorganisms to hydrogen. In the case of using fixed bacteria, compulsory contact of bacteria with an electrode acting as an electron donor is increased and, as a result, denitrification is performed more rapidly. In BERs, organic or inorganic carbon sources can be used to feed microorganisms that are known as heterotrophic and autotrophic hydrogenotrophs, respectively.

THE THEORY OF DENITRIFICATION IN BERs

The expected pathway for the complete reduction of nitrate to nitrogen gas includes four consecutive steps as follows. The oxidation state of nitrogen atoms in each step has been also shown.

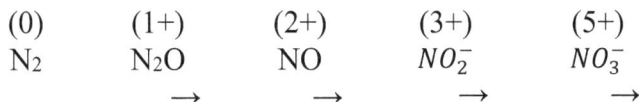

(0)	(1+)	(2+)	(3+)	(5+)
N_2	N_2O	NO	NO_2^-	NO_3^-
	$\rightarrow$	$\rightarrow$	$\rightarrow$	$\rightarrow$

The decomposition of nitrate in a BER system can be shown by equation 7 in the following, which also includes electrolysis of water: Electrolysis of water:

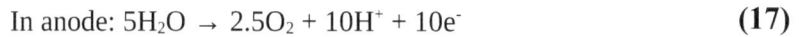

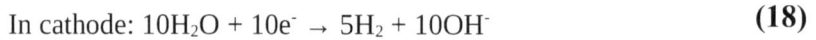

$$\text{In anode: } 5H_2O \rightarrow 2.5O_2 + 10H^+ + 10e^- \tag{17}$$

$$\text{In cathode: } 10H_2O + 10e^- \rightarrow 5H_2 + 10OH^- \tag{18}$$

$$2\,NO_3^- + 2H_2 \rightarrow 2\,NO_2^- + 2H_2O \tag{19}$$

The autotrophic denitrification of nitrate to nitrogen gas has been presented in the following, and the main role of hydrogen gas as an electron donor has been illustrated well:

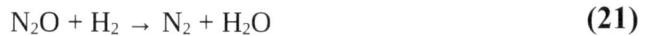

$$2\,NO_2^- + 2H_2 \rightarrow N_2O + H_2O + 2OH^- \tag{20}$$

$$N_2O + H_2 \rightarrow N_2 + H_2O \tag{21}$$

The general reaction of denitrification in a cathode has been summarized as follows:

$$2\,NO_3^- + 6H_2O + 10e^- \rightarrow N_2 + 12OH^- \tag{22}$$

The overall reaction of the bio-electrochemical reactor can be summarized as follows:

$$2\,NO_3^- + H_2O \rightarrow N_2 + 2.5O_2 + 2OH^- \tag{23}$$

Design and Characteristics of BERs

Specifications of electrodes and the design of reactors play a very important role in the rate of denitrification. The main parameters for the production of a bio-electrochemical reactor are material, shape, and number of electrodes and their arrangement inside the reactor as well as the shape of the BER reactor itself. In available reports, various types of metals (stainless steel, titanium, nickel and copper) and carbon materials (granular activated carbon, graphite, carbon fiber) have been used to make cathodes and anodes. The shape of the electrodes and their arrangement have been studied in many studies, and the most commonly used forms in the studies were rods, plates and circles. Parallel arrangement of plate electrodes is the most common electrode system that has many advantages, such as ease of construction and uniformity of electrical current across the entire electrode, but this shape and arrangement of electrode have good efficiency only for high concentrations of contaminants, such as those used for electrochemical reduction of metal ions. In addition to the above-mentioned shapes and arrangements, the porous materials such as filled substrates, mesh or lattice plates, metal plates and metal grids are also used in both forms of plate and circular in studies. The use of several cathodes leads to higher efficiency in the removal of contaminants due to providing a more available surface for attaching denitrifying bacteria and hydrogen production, as well as a larger contact area for contact with the pollutant. In this case, the arrangement of cathodes is of great importance that circular electrodes and parallel plates are used mostly.

MICROBIOLOGICAL ECOLOGY OF BIOLOGICAL DENITRIFI-CATION IN WASTEWATER

Biological denitrification is the sequential reduction of nitrate or nitrite to nitrogen gas through nitric oxide and nitrous oxide gas intermediates. This energy generating a respiratory process is catalyzed by four types of nitrogen reductase, which are placed in sequence; these nitrogen reductases include nitrate reductase (Nar), nitrite reductase (Nir), nitric oxide reductase (Nor) and nitrous oxide reductase (Nos) [5, 27]. In the past century, many studies done in the field of denitrification, the reasons for which can be stated as follows: first, denitrification is one of the main branches of the biogeochemical cycle of nitrogen, which returns reactive nitrogen to the atmosphere and leads to nitrogen balance. The second reason is that denitrification is considered as one of the important processes for achieving Biological Nutrient Removal (BNR) and has been widely used in engineered wastewater treatment systems. The third reason is that biological denitrification can be involved in the global greenhouse gas phenomenon through the release of nitrous oxide (N_2O) gas that is 300 times more powerful for the Global Warming Potential (GWP) than carbon dioxide. Although the denitrification potential is widely observed among bacteria, arches and some eukaryotes (such as fungi), but in natural and engineering ecosystems for sewage treatment, the nitrate reduction is mostly carried out by bacteria. Most of the denitrifying bacteria are arbitrary anaerobic that use nitrogen oxide gas and ions as electron acceptors in the absence of oxygen. The electron donation can be obtained from organic matter (Chemoorganoheterotrophs) or mineral compounds (Chemolithoautotrophic). To date, only a few chemolithoautotrophs have been identified with denitrification ability, while the chemoorganoheterotrophs are available in a wide variety of taxonomic and physiological groups. The phylogenetic diversity of heterotrophic denitrifies, such as their presence in all aqueous and soil environments, is high.

THE FACTORS CONTROLLING THE POPULATION STRUCTURE OF DENITRIFIERS

Carbon Source

Although methanol as the most common carbon resource has been used to intensify denitrification, but in order to exacerbate denitrification, it is necessary to consider other factors such as operation cost, kinetics, and acclimation period [2, 5, 28]. In one system, the carbon source as an organic carbon source and metabolism energy (from different pathways for heterotrophic growth) have a high potential for affecting on the population structure of dominant denitrifying bacteria compared to other factors (electron acceptor and C/N ratio). Specifically, the metabolism of single-carbon compounds (methanol, formate, and methane) is exclusively for methylotrophs, since there is a non-common key enzyme that can oxidize methanol to formaldehyde. This can be the reason for the presence of a more population of denitrifiers consuming carbon source with single-carbon compared to the population of denitrifiers consuming carbon source with multi-carbon.

Methanol

In all methods (diagnostic methods based on culture and molecular methods), the microbial population associated with *methylophilus, paracoccus, methyloversatilis* and *Hyphomicrobium* have been identified in denitrification systems fed with wastewater containing methanol, which specifically belong to beta-proteobacteria. These microbial populations can be classified under compulsory methylotrophs (capable to grow only using C1 (single-carbon) compounds) and arbitrary methylotrophs (capable to grow using single-carbon and multi-carbon compounds). Denitrifying bacteria belonging to the first group (for example, the *hyphomicrobium sp.* strain) have been identified almost in methylotroph systems and those of the last group (*i.e.*, *methyloversatilis* and *paracoccus* spp. strains) have been identified in denitrifier reactors fed with other types of carbon sources. Generally, methylotrophs have limited variation and metabolically differentiate from other non-methylotrophs denitrifying bacteria, so the long initial delay phase (up to several months) in running the methanol-fed denitrification systems is resulted by the enrichment of the methylotrophic population and, consequently, a change in the population structure of the denitrifiers. In fact, bacteria consuming methanol in both lab-scale and full-scale reactors observe only in a small proportion of the denitrifiers, and the growth of other microbial/bacterial populations is possible using by-products derived from the metabolism of methanol or products of the biological decomposition of methanol.

Ethanol

Although many studies have been conducted on the use of methanol as a carbon source of denitrification, the study of the microbial ecological characteristics of denitrifiers using other carbon sources has been few and has recently attracted the attention of many researchers. It has been reported that activated sludge samples that used methanol and ethanol as an organic carbon source were analyzed using Restriction Fragment Length Polymorphism (RFLP) based on nirK and nirS. Based on the reported results, the activated sludge fed with methanol had a higher variety in terms of nirS than the sample fed with ethanol, while nirK was similar for both types of sludge.

Acetate

Some of the bacteria identified in acetate-fed denitrifying systems that use acetate as an organic carbon source are related to the strains of Comamonas, *Acidovorax* and *Thauera* spp. belonging to the family of *Comamonadaceae* and *Rhodocyclaceae*. The microbial diversity of the acetate-fed systems are higher than that of methanol-fed one and the same strains is almost identified in the main sludge sample, which is not fed with any carbon source. Some researchers have observed that the dominance of a particular group of denitrifiers using acetate is possible under the excess substrate and the limited conditions and the availability of an electron acceptor (*e.g.* acetate and nitrite) as a selective factor.

Other Sources of Carbon

In some studies, the population structure of denitrifiers feeding with acetate and a mixture of complex substrates (acetate, ethanol, and pyruvate) has been studied and reported [29 - 31]. The bacterial population identified in both cases was related to alpha, beta, and gamma proteobacteria, among which *Aquaspirillum* was reported the most frequent population (20% of total microbial population); although the *Accumulibacter* (7-3%) and *Azoarcus* (2-13%) had the lowest abundance, but were introduced as key denitrifiers in the case of the use of a mixture of carbon sources (acetate and other carbon sources such as pyruvate and ethanol). *Azoarcus* spp. strain is reported as the main group of bacteria that use acetate and is the only methylotroph that uses acetate and methanol mixture.

Input Wastewater

Industrial and Urban Wastewater

Typically, household wastewater contains 10-40 mg-N/L in the form of ammonia and organic nitrogen, which are converted to nitrate after complete nitrification.

The composition of the industrial wastewater entering the denitrification stage varies greatly depending on the type of the industry, but all industrial wastewaters contain high levels of nitrate and many other types of ions, such as chloride and sulfate. *Azoarcus* and *Aquaspirillum* bacteria are two predominant bacteria, which have been observed in the denitrification systems used in the treatment of industrial and urban wastewaters. Urban landfill leachate is a mixture of high concentration organic and inorganic pollutants including humic acids, ammonia, heavy metals and other mineral salts. Thauera, *Acidovorax*, and *Alcaligenes* are the bacteria identified in the anoxic reactor used to treat the landfill leachates that are rich in nitrate and aromatic compounds.

SALINITY

Some researchers have studied the effect of salinity on the dominance of bacteria population [32 - 34]. The strains of *Halomonas* and *Marinobacterin* spp. in the group of gamma-proteobacteri are among the dominant strains identified by the t-RFLP technique in the denitrifiers acclimated to salinity change in sewage from low to high. These strains have been also identified in the presence of high salinity acetate. Some researchers have identified methanol as the appropriate electrolyte in low salinity (0-3%), and *Azoarcus* and *Methylophaga* strains have been isolated under this condition. *Hyphomicrobium*, *Phyllobacteriacea* and *Paracoccus* strains of the gamma-proteobacteri group have been reported as denitrifier bacteria identified in similar conditions. The predominance of denitrifiers of the Gammaproteobacteria has also been reported in the metallurgy industry saline wastewater containing acetate.

COD/N RATIO

Some researchers have reported a significant relationship between denitrifying bacteria and COD/N ratio. In one anoxic reactor used for treating of urban landfill leachates, it has been observed that denitrification capacity is increased by increasing the COD/N ratio and dominate denitrifiers are changed from autotroph strains of *Thiobacillus* Spp. to heterotroph strains of *Azoarcus* spp.

BIOFILM GROWTH

Higher heterogeneity (for example, substrate concentration gradient, middle and final metabolic product) in biofilm provides the possibility of the presence of different bacteria groups with different metabolic properties. Therefore, there is greater microbial populations in the biofilm compared to the active sludge. In a fluid bed denitrifier reactor, which was used with ethanol as an electron donor, the strain of *Azoarcus* spp. was a dominant denitrifier at first; its population was reduced over time and was replaced by *Dechloromonas*, *Pseudomonas* and

Hydrogenophaga. Since biofilm systems are useful for slow-growing bacteria such as denitrifiers, some nitrogen removal reactors are being fabricated for simultaneous nitrification and denitrification. In the four-stage Bardenpho biofilm process, denitrification mainly occurs in the post denitrification (PD region), and the denitrifier bacteria are related to strains of *hyphomicrobium*, *Rhodopseudomonas* and *Rhodobacter* spp. However, nitrifier bacteria are present in higher parts of PD-biofilm, and their number is low. In the fixed bed and fluid bed denitrifier reactor where methanol was used as a carbon source, *Methylophaga* spp. has been reported as the bacterial strain with the highest frequency and *Hyphomicrobium* spp. has been reported in the second place.

OPERATING CONDITIONS

The operation parameters such as solid retention time (SRT), pH and DO, not only effect on the total removal of nitrogen and denitrification activity, but also effect on the survival and diversity of the dominant microbial population in the process. The results of a study on the effect of mean cell residence time (MCRT) on the microbial population diversity in anoxic regions of the submerged membrane bioreactor denitrification have shown that the microbial diversity of the mentioned regions for short MCRTs (5-0.38 d) has a significant difference with that of long MCRTs (16.7-33.3 d) there is a significant difference. The highest levels of nitrogen removal were also achieved in long MCRTs with fewer strains of bacteria (competitive elimination is dominated in the long MCRT). The denitrification process is sensitive to pH fluctuations, and pH values in the range of 6.5 and 8.5 are appropriate for sludge flocculation. The phylogenetic analysis (based on the cloning) performed on the denitrifiers in a fluid bed reactor has shown that the bacteria population diversity is increased by pH changes from $9 \leq$ to an optimum range of 6.5-7.5. Although the amount of DO is close to zero (0.2-0.5 mg/L) in anoxic denitrification, but aerobic denitrifying bacteria identified in both types of natural and engineered wastewater treatment systems that are able to tolerate DO with the amount of 5-6 mg O_2/L. In an anoxic denitrifier reactor, aerobic denitrifying can be selected/dominated by changing the conditions between oxic and anoxic. Most biological nitrogen removal (BNR) systems are based on biomass recovery between anoxic and aerobic regions, but there is no specific ecological niches for aerobic denitrifying bacteria in wastewater treatment systems.

FACTORS CONTROLLING THE FUNCTION OF BACTERIA AND THE DOMINANT MECHANISMS

The main factors controlling the function of the denitrifying population include a carbon source, pH, temperature, DO and nitrogen oxides. The effects of these parameters on the removal of total nitrogen, denitrification rate and accumulation of by-products studied in different studies. Cellular changes versus the changes for each of these parameters have also been studied using different molecular techniques. In the following, it has been tried to present the effect of each parameter in separate sections:

Carbon Source

The effect of methanol and other electron donors such as acetate, ethanol, and glycerol, which use commonly in denitrification of wastewater, has been investigated on the kinetics of the process and biomass production. In terms of stoichiometry, due to significant energy loss that occurs during methanol assimilation, the methylotroph denitrifying bacteria have less cellular yield (sludge production) than other denitrifiers that grow in the presence of other carbon sources. Because of the limited metabolic capability of mandatory methylotrophs, most denitrifying population acclimated with methanol as a carbon source metamorphic community do not have the ability to use other multi-carbon compounds and *vice versa*. Therefore, acclimation of methylotrophs needs more time and denitrification with methanol will be kinetically inadequate in winter. However, the oxidation state of carbon sources has a low effect on the synthesis of reductases denitrification (such as periplasmic or membranous reductases), the expression of amount of carbon oxidase (such as oxidation of glycerol and dehydrogenation alcohol catalyzing methanol) varies depending on carbon type. As a result, the differences in the denitrification kinetics can be due to an imbalance between the supply of electrons and the rate of its consumption. The accumulation of by-products of the denitrification process, such as nitrite, NO, and N_2O, could be related to competition between four nitrogen reductases for the available electron. The displacement rate (reversibility) of the electron of Nar is higher than Nir; therefore, when the electron donor is sufficiently available, the nitrate is reduced much faster than nitrite. In practice, nitrite accumulation occurs when certain types of carbon sources (such as biodiesel waste) are used or the amount of available carbon is low. In these mentioned cases, the competition between Nar and Nir is very high for electrons. Few studies have been conducted to explain the relationship between carbon and nitrogen metabolism over the denitrification process using various carbon sources. In one study, a good correlation was found between the rate of denitrification (methanol and glycerol have been used as a carbon source) and the level of expression of the alcohol-dehydrogenase genes, and, considering its specificity, it was then used as biomarker to express activity in the denitrification site specific for a type of carbon source.

Temperature and pH

The optimal temperature and pH for conducting denitrification of sewage are 20-30 °C and 7-9, respectively, and the activity of denitrification above and below this range is decreased rapidly and severely. At lower pH levels, the accumulation of middle products of denitrification such as nitrite and nitrous oxide has been observed. The rate of N_2O reduction is much higher dependent on pH than the rate of reduction of nitrate and nitrite in methanol-rich bacteria as a carbon source. In a study, denitrification was performed using the specific strain of *paracoccus*, by decreasing the pH below the optimum range of 6.8, although the denitrification activity of this bacterium has been decreased, but the values of nar H, nirS and nosZ mRNA have not been changed. Most denitrifying bacteria are more sensitive to temperature fluctuations than pH fluctuations. For example, the study of the effect of pH and temperature on the gene expression of the denitrification in *Pseudomonas mandelii* using the reverse transcription polymerase chain reaction (RT-PCR technique showed that the expression of nirS and norB in the pH range of about 8-6 was same, while the cellular growth of the bacterium at 30 °C was much higher than its growth at temperatures of 20 °C and 10 °C.

Dissolved Oxygen (DO)

Oxygen prevents denitrification through the availability of better electron acceptors for the bacteria to produce energy, and results in the accumulation of middle nitrogen products at high concentrations of DO. In practice, in wastewater treatment denitrification systems the oxygen inhibition threshold has been reported to be 0.1 mg O_2/L. Almost the expression and activity of all nitrogen reductases are suppressed in the presence of oxygen. Oxygen has an immediate and reversible inhibitory effect on nitrate respiration and its maximum inhibitory effect results at an oxygen saturation of 0.2%. Nitrite reductase is less sensitive to nitric oxide than oxygen and its inhibition threshold is estimated 2.5 mg O_2/L. Nitric oxide reductase activity is approximately 10 times higher than nitrite reductase, which results in low NO accumulation, which is also toxic to bacteria. NO reductase is the most sensitive enzyme in comparison with other reductases, which results in the accumulation of N_2O transition product in aerobic conditions.

Nitrogen Oxides

In the denitrification process, nitrate and other products, the denitrification intermediates are used by the electron transport chain as the electron acceptor and affect the expression and activity of each of the reductases. In the absence of nitrogen oxides, anaerobic bacteria cannot simulate the acceptable synthesis of four types of reductase. Nitrate, which is preferable to other types of nitrogen oxides as electron receptors, stimulates the synthesis of all of the radicals in the same way. On the other hand, inhibition caused by nitrite has very complex effects on the denitrification process. Research results show that in nitrite inhibitory effects, the inhibitory effects of un-dissociated nitrous acid (HNO_2) are created and its inhibitory threshold concentrations vary based on growth conditions, pH and available carbon. For example, the accumulation of NO and N_2O has been observed over denitrification in the presence of nitrite in feeding bacteria with acetate as carbon source, while it has not been observed in the case of using methanol or ethanol as carbon source. Nitric oxide almost completely inhibits the synthesis of all nitrogen reductases, and usually is available in very low concentrations in bacteria. Nitrous oxide does not inhibit any of the stages of denitrification and even stimulates the synthesis of nitrous oxide reductase.

CONCLUSION

Conventional wastewater treatment techniques found not able to remove the toxic substances, heavy metals, nitrogen, phosphorous *etc.*, effectively or their promotion to obtain the required efficiency is very energy consumable that led to be expensive and not economic.Therefore, an effort to find a new alternative with lower energy consumption was necessary. Bioelectrochemical technologies are considered as a process for the generation of electricity and valuable chemicals from cheap organic compounds along with waste remediation. This process can overcome energy consumption costs, considered highly efficient wastewater treatment and can be nominated as one of the main promising methods. According to the unique characteristics like green energy generation, it is a candidate as a clean and environmental friendly pathway to remediate wastewater in the future.

CONSENT FOR PUBLICATION

Not applicable.

REFERENCES

[1]   Hoseinzadeh E, Rezaee A, Farzadkia M. Nitrate removal from pharmaceutical wastewater using microbial electrochemical system supplied through low frequency-low voltage alternating electric current. Bioelectrochemistry 2018; 120: 49-56.
        [http://dx.doi.org/10.1016/j.bioelechem.2017.11.008] [PMID: 29175273]

[2]   Hoseinzadeh E, Rezaee A, Farzadkia M. Enhanced biological nitrate removal by alternating electric current bioelectrical reactor: Selectivity and mechanism. J Mol Liq 2017; 246: 93-102.
        [http://dx.doi.org/10.1016/j.molliq.2017.09.048]

[3]   Hoseinzadeh E, Rezaee A, Farzadkia M. Low frequency-low voltage alternating electric current-induced anoxic granulation in biofilm-electrode reactor: A study of granule properties. Process Biochem 2017; 56: 154-62.
        [http://dx.doi.org/10.1016/j.procbio.2017.02.019]

[4]   Wan D, Liu H, Qu J, Lei P. Bio-electrochemical denitrification by a novel proton-exchange membrane electrodialysis system-a batch mode study. J Chem Technol Biotechnol 2010; 85(11): 1540-6.
        [http://dx.doi.org/10.1002/jctb.2465]

[5]   Duca M, Koper MTM. Powering denitrification: the perspectives of electrocatalytic nitrate reduction. Energy Environ Sci 2012; 5(12): 9726-42.
[http://dx.doi.org/10.1039/c2ee23062c]

[6]   Saratale GD, Saratale RG, Shahid MK, *et al.* A comprehensive overview on electro-active biofilms, role of exoelectrogens and their microbial niches in microbial fuel cells (MFCs). Chemosphere 2017; 178: 534-47.
[http://dx.doi.org/10.1016/j.chemosphere.2017.03.066] [PMID: 28351012]

[7]   Gude VG. Wastewater treatment in microbial fuel cells–an overview. J Clean Prod 2016; 122: 287-307.
[http://dx.doi.org/10.1016/j.jclepro.2016.02.022]

[8]   Saba B, Christy AD, Yu Z. Sustainable power generation from bacterio-algal microbial fuel cells (MFCs): An overview. Renew Sustain Energy Rev 2017; 73: 75-84.
[http://dx.doi.org/10.1016/j.rser.2017.01.115]

[9]   Hoseinzadeh E. Study of nitrate removal efficiency in bio-electrochemical process with alternating current (low voltage-very low frequency) with Ibuprofen as an organic carbon source (Unpublished doctoral dissertation). Tehran, Iran: Tarbiat Modares University 2018.

[10]   Mook W, Chakrabarti M, Aroua M, *et al.* Removal of total ammonia nitrogen (TAN), nitrate and total organic carbon (TOC) from aquaculture wastewater using electrochemical technology: A review. Desalination 2012; 285: 1-13.
[http://dx.doi.org/10.1016/j.desal.2011.09.029]

[11]   Jafary T, Wan Daud WR, Ghasemi M, Abu Bakar Mn, Sedighi M, HongKim B, *et al.* Clean hydrogen production in a full biological microbial electrolysis cell. Int J Hydrogen Energ 2018. (Accepted/In press).
[http://dx.doi.org/10.1016/j.ijhydene.2018.01.010]

[12]   Karthikeyan R, Cheng KY, Selvam A, Bose A, Wong JWC. Bioelectrohydrogenesis and inhibition of methanogenic activity in microbial electrolysis cells - A review. Biotechnol Adv 2017; 35(6): 758-71.
[http://dx.doi.org/10.1016/j.biotechadv.2017.07.004] [PMID: 28709875]

[13]   Yu Z, Leng X, Zhao S, *et al.* A review on the applications of microbial electrolysis cells in anaerobic digestion. Bioresour Technol 2018; 255: 340-8.
[http://dx.doi.org/10.1016/j.biortech.2018.02.003] [PMID: 29444757]

[14]   Liu D, Lei L, Yang B, Yu Q, Li Z. Direct electron transfer from electrode to electrochemically active bacteria in a bioelectrochemical dechlorination system. Bioresour Technol 2013; 148: 9-14.
[http://dx.doi.org/10.1016/j.biortech.2013.08.108] [PMID: 24035815]

[15]   Wu Y, Li F, Liu T, Han R, Luo X. pH dependence of quinone-mediated extracellular electron transfer in a bioelectrochemical system. Electrochim Acta 2016; 213: 408-15.
[http://dx.doi.org/10.1016/j.electacta.2016.07.122]

[16]   Yang Y, Xu M, Guo J, Sun G. Bacterial extracellular electron transfer in bioelectrochemical systems. Process Biochem 2012; 47(12): 1707-14.
[http://dx.doi.org/10.1016/j.procbio.2012.07.032]

[17]   Zhang S, You J, Kennes C, *et al.* Current advances of VOCs degradation by bioelectrochemical systems: A review. Chem Eng J 2018; 334: 2625-37.
[http://dx.doi.org/10.1016/j.cej.2017.11.014]

[18]   Kim B, An J, Fapyane D, Chang IS. Bioelectronic platforms for optimal bio-anode of bio-electrochemical systems: From nano- to macro scopes. Bioresour Technol 2015; 195: 2-13.
[http://dx.doi.org/10.1016/j.biortech.2015.06.061] [PMID: 26122091]

[19]   Mook WT, Aroua MKT, Chakrabarti MH, *et al.* A review on the effect of bio-electrodes on denitrification and organic matter removal processes in bio-electrochemical systems. J Ind Eng Chem 2013; 19(1): 1-13.

[http://dx.doi.org/10.1016/j.jiec.2012.07.004]

[20]   Kokko M, Epple S, Gescher J, Kerzenmacher S. Effects of wastewater constituents and operational conditions on the composition and dynamics of anodic microbial communities in bioelectrochemical systems. Bioresour Technol 2018; 258: 376-89.
[http://dx.doi.org/10.1016/j.biortech.2018.01.090] [PMID: 29548640]

[21]   Hoseinzadeh E, Rezaee A, Fazeli S. Electrochemical denitrification using carbon cloth as an efficient anode. Desalin Water Treat 2017; 97: 244-50.
[http://dx.doi.org/10.5004/dwt.201x.21692]

[22]   Hoseinzadeh E, Rezaee A, Fazeli S, *et al.* Changing free residual chlorine, Hardness and Alkalinity during electrochemical denitrification process. Pajouhan Sci J 2017; 15(4): 1-9.

[23]   Ghafari S, Hasan M, Aroua MK. Bio-electrochemical removal of nitrate from water and wastewater--a review. Bioresour Technol 2008; 99(10): 3965-74.
[http://dx.doi.org/10.1016/j.biortech.2007.05.026] [PMID: 17600700]

[24]   Jiang Y, Su M, Zhang Y, *et al.* Bioelectrochemical systems for simultaneously production of methane and acetate from carbon dioxide at relatively high rate. Int J Hydrogen Energy 2013; 38(8): 3497-502.
[http://dx.doi.org/10.1016/j.ijhydene.2012.12.107]

[25]   Park Y, Park S, Nguyen VK, *et al.* Complete nitrogen removal by simultaneous nitrification and denitrification in flat-panel air-cathode microbial fuel cells treating domestic wastewater. Chem Eng J 2017; 316: 673-9.
[http://dx.doi.org/10.1016/j.cej.2017.02.005]

[26]   Virdis B, Read ST, Rabaey K, Rozendal RA, Yuan Z, Keller J. Biofilm stratification during simultaneous nitrification and denitrification (SND) at a biocathode. Bioresour Technol 2011; 102(1): 334-41.
[http://dx.doi.org/10.1016/j.biortech.2010.06.155] [PMID: 20656477]

[27]   Lu H, Chandran K, Stensel D. Microbial ecology of denitrification in biological wastewater treatment. Water Res 2014; 64: 237-54.
[http://dx.doi.org/10.1016/j.watres.2014.06.042] [PMID: 25078442]

[28]   Hoseinzadeh E, Rezaee A, Hossini H. Biological nitrogen removal in moving bed biofilm reactor using ibuprofen as carbon source. Water Air Soil Pollut 2016; 227(2): 46-59.
[http://dx.doi.org/10.1007/s11270-015-2690-1]

[29]   Bomberg M, Mäkinen J, Salo M, Arnold M. Microbial Community Structure and Functions in Ethanol-Fed Sulfate Removal Bioreactors for Treatment of Mine Water. Microorganisms 2017; 5(3): 61-78.
[http://dx.doi.org/10.3390/microorganisms5030061] [PMID: 28930182]

[30]   Morgan-Sagastume F, Nielsen JL, Nielsen PH. Substrate-dependent denitrification of abundant probe-defined denitrifying bacteria in activated sludge. FEMS Microbiol Ecol 2008; 66(2): 447-61.
[http://dx.doi.org/10.1111/j.1574-6941.2008.00571.x] [PMID: 18811652]

[31]   Timmers RA, Rothballer M, Strik DPBTB, *et al.* Microbial community structure elucidates performance of Glyceria maxima plant microbial fuel cell. Appl Microbiol Biotechnol 2012; 94(2): 537-48.
[http://dx.doi.org/10.1007/s00253-012-3894-6] [PMID: 22361855]

[32]   Bassin JP, Pronk M, Muyzer G, Kleerebezem R, Dezotti M, van Loosdrecht MC. Effect of elevated salt concentrations on the aerobic granular sludge process: linking microbial activity with microbial community structure. Appl Environ Microbiol 2011; 77(22): 7942-53.
[http://dx.doi.org/10.1128/AEM.05016-11] [PMID: 21926194]

[33]   Cui Y-W, Zhang H-Y, Ding J-R, Peng Y-Z. The effects of salinity on nitrification using halophilic nitrifiers in a Sequencing Batch Reactor treating hypersaline wastewater. Sci Rep 2016; 6: 24825-36.
[http://dx.doi.org/10.1038/srep24825] [PMID: 27109617]

[34]   Mrkonjic Fuka M, Gesche Braker SH, Philippot L. Molecular Tools to Assess the Diversity and Density of Denitrifying Bacteria in Their Habitats A2 - Bothe, Hermann. Biology of the Nitrogen Cycle. Amsterdam: Elsevier 2007; pp. 313-30.
[http://dx.doi.org/10.1016/B978-044452857-5.50021-7]

SUBJECT INDEX

A

Acceptor 30, 33, 59, 95, 96, 115, 118, 120, 127
  common electron 59
  external electron 118
  main electron 33
  reversible electron 120
  solid electron 127
  terminal electron 30, 115
Acetate 29, 49, 59, 63, 69, 87, 88, 89, 115, 116, 127, 134, 135, 137, 139
Acetate-fed denitrifying systems 134
Acid 29, 48, 101, 121, 122
  acetic 29, 101
  -carbonate buffers 121
  nitric 122
  produced 48
Acinetobacter junii YB 75, 76
Activated 9, 15, 16
  carbon 9, 15, 16
  carbon production 15, 16
Activated sludge 28, 50, 53, 61, 71, 74, 87, 88, 134
  nitrification 71
Activity 15, 33, 34, 35, 37, 56, 60, 62, 66, 87, 115, 120, 121, 137, 138
  bio-electro catalytic 115
  bioelectrochemical 87
  electrochemical 120
  microbial 66, 120, 121
  nitric oxide reductase 138
  of anaerobic Anammox bacteria 37
Adsorbate 15, 16
Adsorbents 15, 16
Adsorption processes 15
Advanced oxidative processes (AOPs) 14, 15, 16
Aerobic 29, 36, 37, 38, 46, 47, 55, 68, 72, 73, 74, 75, 77, 138
  bacteria 29, 37, 38
  conditions 36, 47, 55, 68, 138

deammonification 46, 72, 73, 74
deammonification process 72, 73
denitrification 36, 73, 74, 75
denitrification route 75
denitrifiers 74
nitrite oxidizers 77
Aerobic nitrifiers 60, 68, 73
  conventional 68, 73
Air saturation 65
Aldehyde dehydrogenase 96
Alkalinity consumption 54, 76, 77
Alumina NPs 10
Ammonia 2, 4, 5, 8, 9, 10, 11, 13, 14, 27, 28, 29, 33, 34, 35, 36, 37, 47, 48, 51, 52, 53, 54, 55, 56, 58, 61, 62, 66, 69, 70, 71, 73, 74, 77, 78, 90, 122, 126, 127, 128, 129, 134, 135
  adsorption 8
  air 8
  concentrations 35, 55
  nitrogen 55
  nitrogen removal 62
  -oxidizing bacteria (AOB) 36, 37, 47, 48, 51, 52, 53, 54, 55, 56, 61, 66, 69, 73
  removal efficiency 5
Ammonia oxidation 28, 35, 65
  anaerobic 65
Ammonium 2, 12, 26, 27, 33, 37, 38, 46, 47, 51, 52, 53, 54, 57, 58, 59, 60, 62, 63, 64, 66, 68, 75, 76, 78, 127, 128, 129, 130
  converting 128, 129
  nitrite mixture 76
  oxidize 57, 58, 75
  ratio of 58, 68
  removal of 127, 129
Ammonium concentration 51, 75, 78, 129
  highest 78
  in sewage 129
  influent 51
  initial 75
Ammonium oxidation 36, 47, 54, 59, 66, 73, 128, 129, 130

Edris Hoseinzadeh

Dr. Edris Hoseinzadeh received his PhD from the Tarbiat Modares University (TMU) in Environmental Health Engineering. He is presently working as a faculty member at the Saveh University of medical sciences. Prior to this, he was a faculty member of Environmental Health Engineering department at the Lorestan University of Medical sciences (LUMS). During this time in LUMS, he was the Co-chairman of the department and Nutritional Health Research Center (NHRC). He established his own scientific journal in 2016, HOZAN Journal of Environmental Sciences, and he is an editor in chief of the journal. He is active in environmental-related research having 4 patents and over 70 publications in national and international refereed journals and has also authored six books about health, environmental analysis, disinfectants and electrochemical process for water and wastewater treatment. He received the best young researcher in 2012- 2013 from multi academic centers (Hamedan University of Medical sciences, LUMS, Mazandaran University of Medical Sciences and Iranian Association of Environmental Health (IAEH)). He received PhD scholarship from Iran's National Elites Foundation during 2016 and 2018.

www.ingramcontent.com/pod-product-compliance
Lightning Source LLC
Chambersburg PA
CBHW041708210326

41598CB00007B/569